Patent Markets in the Global Knowledge Economy

The development of patent markets should allow for a better circulation of knowledge and a more efficient allocation of technologies at a global level. However, the beneficial role of patents has recently come under scrutiny by those favouring 'open' innovation, and important questions have been asked, namely: How can we estimate the value of patents? How do we ensure matching between supply and demand for such specific goods? Can these markets be competitive? Can we create a financial market for intellectual property rights?

In this edited book, a team of authors addresses these key questions to bring readers up to date with current debates about the role of patents in a global economy. They draw on recent developments in economic analysis but also ground the discussion with the basics of patent and knowledge economics. Striking a good balance between institutional analysis, theory and empirical evidence, the book will appeal to a broad readership of academics, students and practitioners.

THIERRY MADIÈS is Professor of Economics at the University of Fribourg, Switzerland. He previously served as an economic advisor on the Council of Economic Analysis to the French Prime Minister. Professor Madiès' field of research is at the crossroads between public economics and public policy, regional economics and the economics of innovation.

DOMINIQUE GUELLEC is Head of the Country Studies and Outlook Division and Directorate for Science, Technology and Industry at the Organisation for Economic Co-operation and Development (OECD). He formerly held the post of Chief Economist at the European Patent Office (EPO) (2004–2005).

JEAN-CLAUDE PRAGER is the Chief Economist of the Greater Paris Corporation (a state-owned company in charge of the Greater Paris metropolitan area project). Prior to this, he was Director of the French Agency for the Diffusion of Technological Information (ADIT).

Patent Markets in the Global Knowledge Economy

Theory, Empirics and Public Policy
Implications

Edited by

THIERRY MADIÈS
Université de Fribourg, Switzerland

DOMINIQUE GUELLEC
*The Organisation for Economic Co-operation and Development
(OECD), Paris*

JEAN-CLAUDE PRAGER
Société du Grand Paris

CAMBRIDGE
UNIVERSITY PRESS

CAMBRIDGE
UNIVERSITY PRESS

University Printing House, Cambridge CB2 8BS, United Kingdom

Published in the United States of America by Cambridge University Press, New York

Cambridge University Press is part of the University of Cambridge.

It furthers the University's mission by disseminating knowledge in the pursuit of education, learning and research at the highest international levels of excellence.

www.cambridge.org
Information on this title: www.cambridge.org/9781107047105

© Cambridge University Press 2014

First published 2014

Printed in the United Kingdom by Clays, St Ives plc

A catalogue record for this publication is available from the British Library

Library of Congress Cataloguing in Publication data
Patent markets in the global knowledge economy : theory, empirics and public policy implications / [edited by] Thierry Madiès, Dominique Guellec, Jean-Claude Prager.
     pages   cm
Includes bibliographical references and index.
ISBN 978-1-107-04710-5 (Hardback)
1. Knowledge management.   2. Intellectual property.   3. Patents.   I. Madiès, Thierry.   II. Guellec, Dominique.   III. Prager, Jean-Claude.
HD30.2.P3764 2013
382'.45608–dc23   2013021188

ISBN 978-1-107-04710-5 Hardback

Contents

The colour plates are situated between pp. 144 and 145

Figures

Tables

Boxes

Contributors

MARTIN A. BADER is Managing Partner of the innovation and intellectual property management advisory group BGW AG, which is a spin-off from the Institute of Technology Management at the University of St Gallen, Switzerland. At the latter he is also Head of the Competence Centre in Intellectual Property Management. He is a European and Swiss Patent Attorney and holds a Master's degree in electrical engineering. Previously, he was Vice President and Chief Intellectual Property Counsel at Infineon Technologies, a leading high-tech semiconductor company worldwide that was carved out from Siemens. He is a visiting professor in several executive education programmes and has been key speaker in numerous expert groups and conferences on innovation and intellectual property management, e.g. for AIPPI, CEIPI, CIP, EC, EPO, IPI, LES, MCI, OECD, VPP, WEF and WIPO. Mr Bader has recently been selected by *iam* magazine as one of the world's leading 300 intellectual property strategists.

MARC BAUDRY is Professor of Economics at the University of Paris Ouest Nanterre la Defense. He was previously Professor at the University of Nantes and was formerly Assistant Professor at the University of Rennes. His research was initially in the field of environmental and resource economics with a special emphasis on the real option analysis of innovation in less polluting technologies and technological change induced by environmental policy. He gradually expanded his work to the study of patents. He has conducted empirical studies on the optimal design of patents. He more specifically made scholarly contributions to the patent renewal approach to patent valuation. He also worked on the measurement of the efficiency of R&D expenditure on the basis of patent counts. His current work deals with the study of the feasibility of a patent rating system, and also investigating how 'green patents' are valued in patent portfolios by investors. He has published articles in international peer-reviewed journals both in the field of environmental and resource economics and in the field of industrial economics.

FRÉDÉRIC CAILLAUD, MD PHD, is Head of the Licensing and Business Development Department at L'Oréal. He is an expert in licensing matters, partnership negotiations, patent monetization, patent analytics and rating systems. He regularly advises French and European policy-makers on open innovation, strategic management of patents and intellectual property monetization.

RENÉ CARRAZ is a junior economist at the OECD. In 2011–12, he has been involved as a consultant for the OECD with the Steering Group on the Governance of International Co-operation of Science, Technology and Innovation for Global Challenges (STIG). Before joining the OECD, Dr Carraz was a research fellow at the department of Managerial Economics, Strategy and Innovation (MSI), University of Leuven, where he was involved in a project that aims at measuring 'radical innovations'. He holds Master's degrees from both Kyoto and Strasbourg universities, and a PhD in economics from Strasbourg University (2011). Dr Carraz teaches and does research on science, technology and innovation. His recent work includes studies of university–industry linkages in Japan and Asia, the effects of patenting on faculty members and the role of patents and publications in university strategy.

DAVID ENCAOUA is Professor of Economics at University Paris I Panthéon Sorbonne, Emeritus since 2009. His research is conducted in two teams: Centre d'Economie de la Sorbonne (CES) and the Paris School of Economics (PSE). He has held various academic positions in different universities (visiting professor, scientific direction of a research laboratory, doctoral school) and institutions (OECD, Ministry of Economy and Finance, CNRS). He has also held an editorial responsibility in various economic journals. His previous research focused on different areas of microeconomic analysis and public policy, including the economics of innovation, industrial organization theory, the economics of science, the intellectual property system, and the European competition and technology policies. He has authored more than fifty articles and six books on these topics. His most recent research focuses on issues related to the implementation of patents, especially when their validity is a priori uncertain, through licensing agreements or through their inclusion as essential patents in technological standards.

OLIVER GASSMANN is Professor of Technology Management at the University of St Gallen, Switzerland and Director of the Institute of Technology Management. After completing his PhD in 1996, he was leading research and advanced development of Schindler Corporation, headquartered in Ebikon, Switzerland. Dr Gassmann has published in leading journals such as *Research Policy, R&D Management, International Journal of Technology Management, Journal of World Business, European Management Journal, IEEE Transactions on Engineering Management, Journal of Technology Transfer and Harvard Business Manager*. At the core of his research is the pervading question of how companies innovate and profit from innovation. Thus, he is dedicated to discovering new approaches to the management of technology and innovation that contribute to firms' competitive advantage.

DOMINIQUE GUELLEC is Senior Economist at the OECD, working on innovation policies and economic growth. From 2004 to 2005, Mr Guellec was Chief Economist of the European Patent Office (Munich). He has authored several books and many articles on patents, innovation and economic growth. His (co-) publications in English include *The Economics of the European Patent System* (2007) and *From R&D to Productivity Growth: The Sources of Knowledge Spillovers and their Interaction* (2004).

YUKO HARAYAMA is Executive Member of the Council for Science and Technology Policy (CSTP) in Japan. Prior to joining the CSTP, she was professor in the Department of Management Science & Technology at Tohoku University Graduate School of Engineering. She has also served as Deputy Director at the OECD Directorate for Science, Technology and Industry, while contributing as Editor-in-Chief of the *Journal of the Intellectual Property Association of Japan*. Her research and publications are mostly in the fields of science, technology and innovation policies.

RÉMI LALLEMENT has been since 2006 an analyst in the Economy and Finance Department of the Centre d'analyse stratégique (CAS, Paris), a research-based advisory institution under the authority of the French prime minister. Previously, he was Project Manager in the Department of Technological and Industrial Development of the Commissariat général du Plan (CGP), the predecessor organization to the CAS. Before that, he was guest scientist at the Institut für Wirtschaftsforschung Halle (IWH) – one of the main German institutes for economic research – and research fellow at the France-based Centre for Information and Research on Contemporary Germany (CIRAC), at the time of German unification. He holds a PhD from the University of Toulouse, after having studied economics at the University Paris 1 Panthéon-Sorbonne. His publications mostly deal with innovation economics and structural policy, as well as with international economics.

FLORIAN LIEGLER is Research Associate at the Institute of Technology Management, University of St Gallen (ITEM-HSG). Prior to this he worked at Sal. Oppenheim as Associate in Investment Banking – Mergers & Acquisitions. His prior work experience includes other investment banking firms such as Raiffeisen Investment AG and the Royal Bank of Scotland. Mr Liegler holds a Master's degree in economics from Vienna University of Economics and Business Administration and is currently enrolled in the University of St Gallen's PhD programme in business innovation.

THIERRY MADIÈS is Professor of Economics at the University of Fribourg in Switzerland. He was a member of the Council of Economic Analysis of the French Prime Minister and a consultant for the World Bank. His work focuses on public finance, international economics and economics of innovation. He has published extensively in peer-reviewed journals, and directed two reports for the Council of Economic Analysis, on innovation and regional development (2010) and on patent markets (2012).

YANN MÉNIÈRE is Assistant Professor of Economics at MINES ParisTech and heads the Mines-Telecom Chair on 'IP and Markets for Technology'. His research and expertise relate to the economics of innovation, competition and intellectual property. Besides his publications in academic journals, he co-authored a textbook on *The Economics of Patents and Copyright* (2004).

ICHIRO NAKAYAMA is a Professor in the School of Law, Kokugakuin University where he has been teaching intellectual property law courses. Mr Nakayama joined Kokugakuin University in 2009 after serving as an Associate Professor in the School of Law at Shinshu University from 2005 to 2009. Prior to his academic career, Mr Nakayama spent many years in the government of Japan. His last position in government was a Deputy Councillor in the Secretariat of Intellectual Property Strategy Headquarters, Cabinet Secretariat from 2003 to 2005. Mr Nakayama originally joined the Ministry of International Trade and Industry (MITI) in 1989. Then he worked in various fields including patent law reform, regulatory reform in energy industries, and defense technology cooperation. Mr Nakayama holds a Bachelor of Law degree from the University of Tokyo, and Master's degrees from the University of Washington and from Columbia University. He has published a number of articles and co-authored a book in the field of patent law.

ANNE PERROT joined MAPP, an economic consultancy, as a partner in 2012. She was previously the Vice President of the French Competition Authority (2004–2012). She belonged to several public think-tanks on competition policy, such as the Economic Advisory Group on Competition Policy (before the European Commission) and the 'group of experts' of the French energy regulator (CRE). She also was a member of several commissions appointed by the French government on subjects such as telecoms deregulation. Dr Perrot is a Professor of Economics at University of Paris I (Sorbonne) and ENSAE. She has also taught at the Paris School of Economics and the Brussels School of Competition. She earned a PhD in mathematics from the University of Paris VI and a PhD in Economics from the University of Paris I.

JEAN-CLAUDE PRAGER has been Chief Economist of the Greater Paris Corporation (the state-owned company in charge of the Greater Paris development project) since September 2010 and expert in innovation policies at the DG Regio. Before this appointment, he was executive manager of the French public agency of technological information. Previously, he had served in French government for three decades. He is the author of numerous papers and reports, and his most recent published books are *18 leçons sur la politique économique*, written with F. Villeroy de Galhau (2nd edn, 2006) and *The Unequal Development of Regions*, with J.-F. Thisse (2012). He has been associate or visiting professor in numerous universities.

ANNE YVRANDE-BILLON has been Associate Professor in Economics at University of Paris I (Sorbonne) since 2003 and senior economist at the French Competition Authority since 2011. Prior to joining the French Competition Authority, she worked at the French Council of Economic Analysis, an independent advisory body reporting to the prime minister. She holds a PhD in economics from the University of Paris I. Her research and publications are mostly in the fields of economics of contracts and organizations, and regulation of utilities, with a particular interest in the public transport sector.

Introduction

DOMINIQUE GUELLEC, THIERRY MADIÈS
AND JEAN-CLAUDE PRAGER

Patents are particular goods, resulting from a legal construction. They guarantee temporary exclusivity rights to the use of an invention to their holder. By offering rights and incentives to investors and inventors, a patent ensures, in a market economy, the decentralization of innovative investment decisions. The cost for society is the temporary exclusion of third parties from the use of the protected invention. Because it restricts both industrial and intellectual use, this social cost is higher when innovation processes are cumulative, i.e. when inventions pile on top of each other.

Patent markets facilitate transfers of this hybrid right and allow us to circumvent the traditional arbitrage between, on one hand, the exclusivity guarantee given to the inventor to encourage investment and, on the other hand, the need of not excluding (and even including) potential users of this invention. When markets function efficiently they can improve the availability of inventions for those that can use them to create value. Furthermore, efficient markets can increase the range of opportunities for these inventions, and finally reduce costs to access such inventions. Consequently, we observe an increase in market transactions on patents and their underlying inventions.

Markets are encouraging this trend. They allow inventions to circulate better between economic agents and their price to be determined, which improves the allocation of inventions in the economy, under certain conditions. The facilitated diffusion of technologies is of great importance. Indeed, technologies can increase productivity by allowing companies to be more efficient, which is particularly important in inventive activities. Improvements in technology diffusion can take the following channels: a deeper division of research, a facilitated access to sources of knowledge to practice the so-called 'open' modes

The authors wish to thank Simon Lapointe, research assistant in the Department of Economics, University of Fribourg, for his help with this chapter.

1

of innovation, and, finally, the emergence of new funding methods for investment in research, i.e. capital may be placed directly in creative assets, valued separately from the other assets of the companies.

The present book analyses the current state of knowledge available to assess the importance of patent markets and to identify the patent strategies of firms and public research organizations.

Chapter 1, 'The market for patents: actors, workings and recent trends', by D. Guellec and Y. Ménière, describes the current state of the market for patents (in the form of licenses or asset transfers), and highlights the principal characteristics of these markets and their actors. The rapid development of patent exchanges in recent years reflects the development of the knowledge economy in general. The separation between proper inventions and assets that allow their economic implementation (physical capital, commercial infrastructure, etc.) is indeed a major trend of the beginning of the twenty-first century. The authors highlight the variety of patent practices regarding the licensing or sale of patents. They oppose in particular 'ex post' enforcement practices (simply aiming at regularizing, through negotiation or litigation, the exploitation of the patented invention by third parties) and the 'ex ante' commercialization of the patented inventions with a view to enabling their exploitation by third parties.

They show that transaction costs associated with these purely private exchange types, on a case-by-case basis and over the counter, constitute an important brake on development of these exchange types. In this context, some innovating instruments have recently been suggested to reduce the costs of these transfers, by operating them at a greater scale and in a more structured fashion. Numerous operators thus try to find economies of scale to ensure the capture of a portion of those productivity gains. This chapter provides a taxonomy of the principal market types (transaction and intermediary types) and presents the principal actors currently involved in them.

Chapter 2, 'Strategic intelligence on patents', by F. Caillaud and Y. Ménière, focuses on new tools for mapping patents and assessing their quality. The search for prior art in patent databases is a necessary yet difficult task for all firms and public research organizations engaged in R&D activities. During the last decades, the growing volume of patents has made this task even more challenging. At the same time, the transition towards the open innovation paradigm has further reinforced innovators' needs for strategic intelligence about their rivals and potential partners' R&D activities. These developments put traditional methods for analysing patent information under strong pressure: the human resources and time they require are hardly compatible with the ever-increasing amount of information to be processed. Against this background, new methodologies enabling the statistical exploitation of patent data on a very large scale offer increasingly relevant solutions, thereby paving the way for a deep renewal of the way in which firms elaborate their R&D strategies. In this chapter, the authors use various examples to illustrate how such innovative instruments can produce critical strategic intelligence by enabling the visualization and quality assessment of entire patent portfolios at both the macro (country, sector) and micro (firm or researcher) levels. Besides searching for prior art, they make it possible to quickly assess the strengths and weaknesses of any firm's intellectual property portfolio, anticipate rivals' strategic R&D orientations and identify infringers or potential partners for technology transfers. Accordingly, they confer significant strategic advantages to the few actors who can already afford to use them. In the coming years, their widespread use is likely to bring about major developments in the innovation ecosystem, by introducing transparency in what still remains one of the more complex and opaque facets of the economy.

Chapter 3, 'Microeconomic foundations of patent markets: the role of intermediaries, auctions and centralized markets', by A. Perrot and A. Yvrande-Billon, relies on recent developments in the literature on market design, auction and transaction costs to explain the various institutional arrangements that coexist for technology transfer, the latter ranging from negotiations over the counter to auctions through hybrid mechanisms, such as joint ventures or cross-licensing. The introduction of market mechanisms to coordinate patents exchanges has not eradicated the other modes of knowledge exchange. While most of the companies surveyed report a strong growth in revenues generated by patent trading and licensing, the majority also highlights the inadequacy of their licensing activity relative to their expectations. By the same token, the few experiments of patent auctions that have recently been organized proved not to be very successful. These facts suggest that market transactions on such particular goods as patents are far from being a 'one-size-fits-all' solution. Does it mean that patents do not easily lend themselves to market exchanges? Or is it due to failures of the existing modes of market coordination? In other words, what are the institutional and structural obstacles to the development of markets for patents? The purpose of this chapter is precisely to provide some answers to these questions using the most recent developments in microeconomic theory.

Chapter 4, 'Structuring the market for intellectual property rights: lessons from financial markets', by O. Gassmann, M. A. Bader and F. Liegler, illustrates how intellectual property rights (IPR) have become a valuable economic commodity in the knowledge economy, gaining in importance as a strategic competitive advantage. Access to IPR is crucial for companies that wish to develop or expand their product range. This raises the question of the optimal allocation of IPR. Today, companies and research organizations already trade and license patents. The sale or licensing of patents to third parties increases innovation and technology transfer, generates economic value and provides access to capital. However, the market lacks transparency, and uncertainty as to the quality and value of patents and technology drives up transaction costs. Trading IPR in a more formal way could facilitate a more efficient allocation process through improved transparency and more accurate pricing mechanisms, thereby affording the greater transactional certainty that the market needs. In order to move towards a more efficient, organized IPR market, and in the interests of fostering trade and engaging with investors, the chapter conceptualizes a market model and identifies potential products to be traded.

In Chapter 5, 'Valuation and rating methods for patents and patent portfolios', M. Baudry deals with valuation and rating methods for patents and patent portfolios. Valuation and rating methods for patents are intended to make it easier for economic agents to discriminate efficiently among a large set of patents and to early detect the more valuable ones. The econometric literature proposes indirect assessment methods based on observable and objectively measurable characteristics of patents, referred to as patent metrics. These metrics are assumed to condition the rent that may be extracted from patents. Assessment methods are said to be indirect in the sense that the level of the rent is not observed but inferred from surveys, observed behaviours or economic results. Most of these methods were initially designed to characterize the overall distribution of values within a population of patents rather than to assess values at the patent level. Indeed, they initially aimed at getting some insights into the pace of innovation at a macroeconomic or sector level and from a more qualitative point of view than simple patent counts. This chapter reviews research articles that adapt them for the purpose of valuation of individual patents or of a portfolio of patents. It is argued that none of the three types of methods described in this chapter is perfect. Methods of the first type, stated value approaches based on an econometric treatment of survey data, are costly to implement and to update and may potentially be subject to declaration biases. As regards the second type, methods based on the valuation by stock markets and more specifically Tobin's Q studies, criticism is based on a background note: in essence, no additional information compared to that available to financiers is produced. Resulting patent scorings cannot therefore be thought of as tools that can help other economic agents than the patent holder to discriminate better among patents. For their part, methods of the third type do not account for the strategic component of the value of patents. Patent renewal methods are particularly illustrative of this third type that gathers revealed value approaches. For the time being, none of the described methods is able to convincingly address the distinction between embodied patents and disembodied patents. Similarly, none of the described methods currently tackles the valuation of a portfolio of patents, though some of them could theoretically do so. Nevertheless, a striking feature of these different methods is that they all point to the same cautious conclusion as regards the feasibility of an automated patent scoring. Indeed, they all conclude that some patent metrics, and more specifically forward citations, have a significant impact on patent value but that the role of patent metrics in explaining the total variance of patents value is rather too limited compared to that of unobserved sources of heterogeneity.

In Chapter 6, 'Dysfunctions of the patent system and their effects on competition', D. Encaoua and T. Madiès argue that the contemporary tensions between patents and competition no longer reside in the traditional trade-off between the exclusionary right given to an inventor to encourage innovation, and the welfare loss induced by the market power associated to this right. Instead, they argue that the three following distortions of the patent system create important conflicts between patents and competition. The first distortion is due to the existence of weak patents. Many patents are granted to applications of bad quality that do not satisfy the usual patentability criteria. This situation increases the uncertainty attached to patents, reduces the credibility of the system and challenges the justification of the patent as a protective mechanism. Second, patents being originally designed in the context of isolated innovations, they are not adapted to the context of sequential or intergenerational innovations, in which an innovation relies on earlier patented inventions. Sequential innovations call for fine delimitations between the rights of successive patent holders, with the patent system operating smoothly as long as each of the successive rights are well defined. However, due in particular to the strategic use of patents to preserve and expand the exclusionary rights by blocking further improvements or unanticipated usage, the patent holders' opportunistic behaviour becomes unavoidable. As in many other circumstances, this opportunistic behaviour is exacerbated by the large extension of the patentable subject matters that appeared in the United States during the 1980s and 1990s. Third, the emergence of complex technologies, corresponding to activities in which the use of a large number of patents from different owners is necessary to create a new product, implies that patent holders will act in a coordinated way. The potential entrants in these complex technologies are then struck by the coordinated behaviour of the patent holders. Examples of such behaviour include the pooling of complementary patents and the licensing of essential patents by the members of a standard setting organization. Often, patents serve to create ambushes or to capture unjustified rents through excessive licence fees, which in turn create barriers to entry for new competitors in the innovation market. Two important consequences of these distortions are derived. First, the resolution of the conflicts between competition and patents cannot rely exclusively on the application of antitrust law. Even if these distortions affect competition in the product, technology and innovation markets, antitrust rules are unable to counter the specific effects rising from distortions of the contemporary patent system. The second consequence is that the existence of these distortions leads to a very expensive judicial implementation of the patent system. The multiplication of conflicts due to a strategic use of patents, particularly in the information and communication technologies, biotechnology and medicine sectors, raises the question whether the legal status of patents is adapted to contemporary technological developments.

In Chapter 7, 'Valorization of public research results and patents: elements of international comparison', R. Lallement underlines the growing concern worldwide about the capacity of publicly funded research institutions to contribute effectively to wealth creation by transferring their results to the business sphere. If this debate is legitimate, several misunderstandings need to be dispelled. The first concerns the notion itself, as a frequent conception of valorization is focused too heavily on the commercialization of IPRs through patenting and licensing. However, it is more realistic to adopt a broader approach which, as reflected by the actual practices of most technology transfer offices (TTOs), corresponds to a much greater variety of tasks, ranging from invention disclosures to contract agreements and creation of spin-offs. Concerning these various activities, the available data tend to be misleading at first glance, suggesting that the USA outperforms other industrialized countries for almost all criteria. But a more cautious international comparison leads to more mixed results, showing that the only indicator for which the USA has a clear leadership in relative terms is the value of licensing revenue. Moreover, structural and institutional factors explain a large part of the performance gaps. Hence, public policies concerning technology transfer and valorization cannot follow a general pattern and must reflect the diversity of missions assigned to different research organizations in question. Yet several general lessons can be learned from the economic analysis and from the experience of diverse countries. One of them is that patent and licensing play a crucial role as incentives in this matter, notably to promote the involvement of academic researchers in close and often long-lasting science–industry partnerships. But they are of varying importance depending on the technological domain considered. Moreover, licensing income varies a lot in time and space according to many factors such as chance or the profile of the respective research organizations. Apart from few exceptions, the vast majority of cases correspond to unprofitable valorization activities, at the level of TTOs. Another result is that size (scientific and human resources) and experience play a major role in explaining a high level of performance. This is why countries like Germany and France have recently created patent and valorization agencies at the regional level. But the idea that these agencies could be self-financing seems illusory. If valorization activities are considered by many experts as a net source of income and therefore as a way to finance academic research, they are in fact a cost factor in most cases. In terms of public welfare, the true rationale of valorization activity should be for governments to promote a wide utilization of results stemming from publicly funded research, not to maximize any financial return, all the more as an excessive commercialism tends to impede public research by undermining the ethics of 'open science'. Where necessary, the need to limit some of these possible negative impacts justifies exploring alternative approaches to practices focusing on systematic patenting, high royalty rates and exclusive licenses.

In Chapter 8, 'Openness, open innovation à la Chesbrough and intellectual property rights', R. Carraz, I. Nakayama and Y. Harayama show that the rise of the open innovation paradigm, a model where the division of innovative labour is widely dispersed, has attracted considerable attention both in academia and in the policy sphere. Indeed, this model entails some considerable changes in the management of innovative activities; in particular, the creation of value requires setting up a business model where firms need to integrate and monetize internal and external knowledge to their organization. In that respect, firms have to build a strategic (intellectual property) management to operate efficiently in this business model. The openness here puts emphasis on the distributive nature of innovation among a wide range of heterogeneous stakeholders rather than an uncontrolled access to it, which may generate new perception and use of IPRs. In contrast to the 'open science' and 'open source software' regimes, the diffusion of knowledge is not unrestricted or uncontrolled but rather its access can be controlled by each stakeholder depending on the strategic goals of the firms, leading to targeted knowledge disclosure. While conventional wisdom puts a focus on exclusivity of a patent right, open innovation *à la* Chesbrough urges company managers to reconsider the role of patents and use them as vehicles for technology transfer in IP markets. It makes it clear that patents are tradable property rights. Keeping it in mind, policy-makers should carefully revisit the institutional design to make sure that technology transfer through IP markets contributes effectively to accelerate innovation and is not obstructed by institutions that have no intention of exploiting patented inventions (the extreme case being 'patent trolls' who aggressively enforce patents against alleged infringers with no intention of manufacturing or marketing the patented invention). Overall, firms need to develop practices to deal with external knowledge flows and to build strategies of knowledge integration tailored to different partners and level of openness. Depending on the circumstances and partners, firms should diffuse their knowledge on an unrestricted basis, build long-term cooperation with different actors, such as universities, or monetize their inventions on IP markets and networks.

1 | Markets for patents: actors, workings and recent trends

DOMINIQUE GUELLEC AND YANN MÉNIÈRE

1.1 Introduction

The growing importance of knowledge flows is strongly backed by anecdotal evidence and widely recognized by practitioners and economists. The apparent expansion of the trade of intellectual property (IP) in general and patents (or titles to patents) in particular is part of this broader trend. It is illustrated by growing volumes, growing stakes, new actors of various types, new policy issues and controversies.

Although companies increasingly seek to divest or acquire patents strategically to strengthen their business, the trade in IP remains inhibited by significant transaction costs. In many cases, patent holders do not have the resources, skills or relationships to identify interested buyers. Moreover, most of them have difficulty in ascertaining the value of their patents. Similarly, most willing patent acquirers do not have enough of the resources and know-how needed to identify the key patents and their proper market prices, to launch and facilitate the negotiations with owners of target patents appropriately and to conclude contracts successfully. For such companies, IP specialist firms now provide various services to support and facilitate patent transactions.

This expansion of patent markets has elicited two opposite views. The first one refers to the trade in IP as trade of technology, which is good as it improves the allocation of knowledge across the economy, hence increasing overall productive efficiency and innovative performance. The second view claims that the trade in patents is often predatory,

This text is largely based on: (1) CAE report (2010) 'Les marchés de brevets dans l'économie de la connaissance'; (2) EC Report (2012) 'Options for an EU Instrument for Patent Valorisation', Report of the Expert Group set up by the European Commission; (3) Guellec, D. and Yanagizawa, T. (2009) 'The Emerging Patent Marketplace', OECD Working Paper. The opinions expressed in this chapter are the sole responsibility of the authors and do not necessarily reflect those of the organizations they work for or that commissioned the aforementioned documents.

a purely legalistic activity, disconnected from real innovation, and thus aimed at capturing rents at the expense of real innovators.

This chapter claims that both views are partly true: patent markets can generate both value creation and rent seeking. As they allow the mobilization of technical knowledge they represent a major opportunity for developed economies, but a number of conditions (notably of a regulatory nature) need to be met for the positive side to dominate. This chapter will not examine in depth what these conditions are, but this double nature of patent markets is an essential conclusion of the following analysis.

The first section will describe patent-based transactions, the second will assess their quantitative importance, and the third will examine market intermediaries.

1.2 Trading patents: conceptual issues

1.2.1 Patent markets encompass a large variety of transactions

Transactions based on patents are of different types. First, the patent itself may be fully transferred by its original owner to a new acquirer. Second, licensing contracts may give the right to use a patented invention under certain conditions. They usually restrict the use of the invention to specific geographic areas or periods of time. The license may also be exclusive (so that the single licensee has a monopoly on the exploitation of the invention) or not (inducing competition between several licensees). Most licensing contracts also include for instance specific conditions for the payment of royalty fees (e.g. the licensee pays the licensor a fixed amount plus a percentage of revenues generated by the patent). 'Cross-licensing' contracts are sometimes an exception in this respect. They aim at enabling the contracting parties to exploit each other's patents in a particular field, and are thus especially frequent in sectors where products embody large numbers of patented components (e.g. information and communications technology (ICT) and car industries). Third and finally, patents may be subject to financial transactions (e.g. securitizations), which allow the holder to monetize his invention without losing control (see Box 1.1).

From an economic perspective, it is insightful to sort these transactions according to whether or not they induce an actual technology transfer between the seller and the acquirer. Technology

Box 1.1 Patent markets in different industries

Trading activities in IP are quite heterogeneous across sectors. Using US data, Anand and Khanna (2000) show for instance that licensing contracts are especially frequent in pharmaceuticals (37.4% of all deals) and in the computers and electronics industries (34.9%).

The incidence of licensing activity (relative to joint ventures and other alliances) is highest in pharmaceuticals and chemicals. These industries are characterized by strong IP rights that provide an effective protection for inventions and the related know-how (Cohen *et al.*, 2000). Licensing contracts typically involve the exclusive transfer of rights over prospective technologies, including manufacturing and marketing rights. They include relatively fewer restrictions than in other industries.

By contrast, patents are weaker appropriation mechanisms in the computers and electronics industries (Cohen *et al.*, 2000). In this context, licensing contracts frequently take place between firms that had prior relationships, and include more restrictions to circumscribe potential imitation. They are usually signed after completion of the research and development phase, and on a non-exclusive basis (Anand & Khanna, 2000).

The computers and electronics industries are also characterized by complex technologies: A particular product (e.g. a computer or a mobile phone) may embody numerous patented components held by different companies (Shapiro, 2001). In this context, IT firms frequently resort to cross-licensing agreements to clear potential infringement liabilities on each other's patent portfolio, and thereby secure freedom to operate (Anand & Khanna, 2000). In recent years, some firms have started purchasing entire patent portfolios to strengthen their bargaining power in such situations. The acquisitions of Nortel's portfolio of 6,000 patents by a consortium led by Apple and Microsoft, and of Motorola Mobility's 17,000 patents by Google (as part of a corporate acquisition) in summer 2012, are striking illustrations of this evolution.

can be defined as a consistent set of information (such as technical specifications and the related know-how) out of which only key elements can be patented. Because these patents include a formal definition of the protected invention, and confer upon their owner the exclusive right to make, use and sell it for a limited period of time, they are a necessary legal support to define, appropriate and monetize intangible technology assets. However, the mere sale or licensing of such rights is usually not sufficient to induce a transfer of the whole underlying technology.

Against this background, transferring technology requires that the buyer/licensor is enabled to exploit effectively the patented invention. It should thus be viewed as a long-term activity which actually exceeds the mere transfer of IP rights. Because the information disclosed in patents seldom suffices to enable effective exploitation, the buyer also needs to acquire the related non-patented know-how. Moreover, technology usually needs to reach the stage of proof-of-concept in order to be suitable for commercialization. Therefore, the seller may have to incur further development costs to support the technology maturation process until that stage is reached.

By contrast, the trade in IP may also proceed from the mere enforcement of patent rights against entities that (allegedly) have already exploited the patented technology. Unlike technology commercialization, the sale or licensing of patents then results from infringement litigation or the threat thereof. Since it proceeds from an infringement, it does not imply any technology transfer and can thus take place in a short-term perspective. Related costs mainly consist of the cost of detecting infringers and the subsequent negotiation and litigation costs.

1.2.2 Social benefits and costs of patent markets

Do patent markets generate economic value? The answer tends to be positive when IP transactions induce the actual transfer of patented technology. It is rather more ambiguous as regards the ex-post enforcement of patent rights.

On the one hand, the trade in patented technology generates value through various mechanisms of technology diffusion and division of labour:

- A company (or university or individual inventor) may be able to invent new products, but lack the skills and assets required to produce and sell them efficiently. Against this background, patent transactions allow the emergence of a division of labour between inventors and manufacturers. Such segmentation of the value chain is a source of collective efficiency gains. The ICT industries are exemplars of this trend, with the involvement of multiple actors in the process.

- Going further in the division of labour, the research process itself can be segmented between actors with different comparative advantages. In the pharmaceutical industry, it involves for example research universities (basic research), biotechnology companies (upstream research), pharmaceutical companies (downstream research), and firms specializing in clinical trials (including phases III and IV). This division of labour should generate productivity gains.
- Intellectual property transactions make it possible to increase the use of each invention. Since knowledge is a non-rival resource its use by an agent does not prohibit simultaneous use by another agent, and it is therefore socially beneficial that each invention be used whenever possible. Whenever licensing agreements allow a greater number of companies to exploit an invention, there is thus a net gain for the society (assuming the total production of these companies exceeds that of a single producer).
- More specifically, patent transactions expand the range of technologies that industrial companies have access to, enabling them to use inventions developed elsewhere to enhance their productivity.
- Finally, patent transactions may be harmful to consumers if they cause restrictions in competition. This is for instance the case whenever a single entity manages to monopolize a market segment by acquiring patents covering different rival inventions. However such anticompetitive practices are forbidden by antitrust authorities (for example in the Licensing Directive of the European Commission, No. 772/2004 of 27 April 2004). And in other cases, licensing agreements, at least non-exclusive ones, tend to increase the degree of downstream competition in the markets for goods embodying the invention, and are thus beneficial to consumers (see Chapter 5 of this book for a more comprehensive discussion of the interface between patents and competition policy).

On the other hand, the mere enforcement of patent rights does not directly contribute to a wider diffusion of technology and knowledge. The ability to enforce patents is nevertheless a necessary condition for the patent system to promote innovation effectively. The benefits that innovators can derive from the exploitation of their inventions indeed depend on the effectiveness of the protection that patents confer against imitators. Their ability to commercialize disembodied technology similarly depends on the legal strength of the patents that are sold or licensed.

In some cases, patent enforcement may even result in perverse effects, especially in technology fields such as ICT where products frequently encompass a large number of overlapping patents. Some non-practising entities – the so-called 'patent trolls' – have specialized in acquiring patents whose legal validity is questionable in order to extract licensing fees from other companies that allegedly exploit the patent. It is indeed cheaper for target companies to pay a settlement fee rather than to challenge the patent in court. Instead of generating incentives to innovate, this form of enforcement is likely to induce extra costs and risks for truly innovating companies. A recent study concludes for instance that litigation cases initiated by non-practising entities between 1997 and 2000 have brought about a total decrease of about \$320 billion of the US stock market value of the sued companies (Bessen *et al.*, 2011). So far trolling has mostly developed in the United States, due to a rather favourable legal system. By contrast it has not developed on a significant scale in other countries.

1.2.3 Patent markets are hampered by high transaction costs

Various types of transaction costs hamper the functioning of patent markets. For clarity of exposition, it is convenient to group them into screening, information and contracting costs.

Screening costs
Patented technologies are highly differentiated and must therefore be valorized on a case-by-case basis. As a first step, patent valorization thus hinges on the seller's ability to find potential buyers for a given technology (and conversely). It is difficult for even an expert to understand the functioning of an invention based on the description disclosed in a published patent, and even harder to identify potential fields of application. A new material can for example be utilized in a multitude of industries. This implies that technology suppliers and acquirers may have difficulty recognizing each other: This informational friction is also one reason for not achieving beneficial transactions.

Against this background, transactions are often initiated through personal and business networks rather than in the open market, for personal ties make it easier to overcome information asymmetries. In a recent survey of the top 1,000 applicants at the European Patent Office (EPO), respondents consider for instance personal networks as the most effective type of intermediary for patent valorization (EC, 2012a). A survey of European innovative small and medium enterprises (SMEs) in turn indicates that personal networks and existing business partners are by far the most effective channels for them to find potential partners, with a rate of success of 35% and 36% respectively (the next most effective channel being patent attorneys with a 24% rate of success) (EC, 2012a).

Information costs

A transaction requires that the parties agree on a price, a value. Estimating a patent's value is a difficult operation, and it is common that estimates vary significantly according to the experts and the methods employed (these methods are presented in more detail in Chapter 6). The main sources of uncertainty are legal risk (Is the patent valid? Is it likely to be invalidated or at least narrowed by a court?), technological uncertainty (Does the invention really work and achieve the expected performance?) and economic uncertainty (Is there a demand for the patented technology? Is there a risk that a competitor might emerge in the meantime with a better invention?). There is no widely recognized method or accepted standard for valuing patents. And each patent is quite different from every other one.

The seller may have more information on the value of an invention than the acquirer (the seller may already have conducted tests, etc.). Although less likely, the opposite can also happen (the potential buyer has a better idea of the market opened by the invention). In any case, this information asymmetry creates the conditions for a 'lemons market': a situation in which mutual distrust between sellers and buyers makes it more difficult to obtain a fair valuation for high-quality items. As a result, the most valuable technologies may be driven out of the market, with only mediocre ones being actually traded.

Contracting costs

These costs include all costs for negotiating the details of property transfer or the licensing or sale agreement as well as the costs for lawyers. The need to include non-patented know-how in the transaction has an inflating effect on contracting costs. The disclosure and transfer of non-patented know-how requires setting up sophisticated contractual frameworks to prevent the uncontrolled leakage of information, while making sure that the technology can be effectively exploited by the buyer. It also has a cost in terms of human resources, insofar as the relevant technical staff must be involved in the transaction.

The lack of structure in markets for patents results in a very imperfect matching of supply and demand, whereby a significant number of transactions that would be beneficial for the parties involved are not realized. A survey carried out in the USA and Canada concludes for instance that firms managed to find a potential licensee for only 25% of the inventions they had sought to license. Negotiations were initiated for 7% of the inventions, and only 4% of them were eventually licensed (Razgaitis, 2004). According to an OECD survey (Zuniga & Guellec, 2009), 24% of European companies and 27% of European companies that own patents report being unable to license all the patents they would like to commercialize on the market for IP. Such a failure rate in transactions, if correctly reported, reflects the inefficiency of the current organization of patent markets.

1.3 Quantitative evidence on patent markets

Market transactions based on patents account for only a small share of the overall trade of knowledge. They encompass two main classes of assets related to patents and technologies: the patent titles themselves, and the rights to use patents (licenses of various types). There is no statistical system providing a reliable and comprehensive estimate of transactions involving such assets. Indeed most of them are private and done over the counter rather than in the marketplace. Companies are moreover reluctant to release any information deemed a 'trade secret'. However, various surveys have been carried out on different facets of the market for patents. Reviewing them makes it possible to collect information on the volume of activity patent markets, and on the origins of the traded assets.

1.3.1 Volume of patent markets

Available figures firstly indicate that about 10% to 15% of patents are licensed (Motohashi, 2008; Nagaoka and Kwon, 2006). Based on a survey of about 7,000 European patents, Gambardella *et al.* (2007)

estimate for instance that 13.4% of patents granted at EPO are licensed (another 50% being used exclusively in-house). In another survey, Kamiyama *et al.* (2006) compared licensing expenses, licensing revenues and R&D spending in a sample of European, American and Japanese firms. They found that both spending on inward licensing and royalty revenues from outward licensing amount to a significantly lower share of R&D spending for European firms than for their US and Japanese counterparts. This suggests that licensing activities are less developed in Europe than in the other two triadic countries.

Recent studies in turn suggest that about 10% of US patents change ownership at least once during their lifetime (Serrano, 2006, 2010). By contrast, statistics for France indicate that only 1.3% of valid patents were traded in 2005. This volume has been increasing rapidly since then, and represented 10% of granted patents in 2009. However, a majority of these changes of ownership are the consequence of merger and acquisitions or intra-group operations rather than the actual sale of patent rights (Ménière *et al.*, 2012). This again suggests that the trade of patents is underdeveloped in Europe as compared with the USA.

Some estimates of the value of the market in IP can be obtained from IP professionals. In the USA, transactions based on patents represented $500 million in 2006 according to practitioners (Monk, 2009). A Silicon Valley intermediary reports having managed a cumulative amount of about $50 million in transactions in a single quarter of 2008. Some recent transactions have even reached a much higher order of magnitude. In summer 2012, a consortium of companies led by Microsoft and Apple paid up to $4.5 billion in cash to acquire Nortel's portfolio of 6,000 patents related to mobile communications. The return on assets may also reach significant values. For instance, in 2003 IBM earned for the first time $1 billion through licenses on its patent (its cumulative revenues from patent licensing can be estimated at about $10 billion since then). This was probably the starting point of a renewal of patent strategies in large companies (although Texas Instruments had accumulated large revenues before IBM through patents licensing).

The balance of payments provides an indicator of international flows of transfer payments associated with IP. Flows (sum of revenue and expenditure) of 'technology balance of payments' in 2008 represented about 0.8% of GDP for OECD countries, against only 0.4%

Table 1.1 *Use of European patents by category of owner*

	Internal use only	License	Unused	Total
Large companies	50	9.2	40.8	100
Medium-sized companies	65.6	10.2	24.2	100
Small companies	55.8	25.8	18.4	100
Private research organizations	16.7	41.6	41.7	100
Public research organizations	21.7	33.3	45	100
Universities	26.2	32.5	41.3	100
Other public organizations	41.7	25	33.3	100
Others	34	29.8	36.2	100
Total	50.5	13.2	36.3	100

Note: Distribution according to the inventor's employer. Number of observations = 7,556.
Source: PatVal (2005).

in 1997. This represents the total global flux of about $120 billion. These figures show a very significant increase. However they include items that are not related to patents, such as transfers of trademarks or copyrights, etc.

1.3.2 Origins of traded intellectual property

The trade of IP is unevenly distributed among patent holders. In a survey carried out in Europe and Japan in 2007, Zuniga and Guellec (2009) found for instance that 20% of European companies and 27% of Japanese companies holding patents have licensed at least one of them. More than half of these licensors license 80% to 100% of their patent portfolio.

As illustrated in Table 1.1, the PatVal survey highlights important differences between categories of owners in this respect (PatVal, 2005). While on average 13.4% of patents are licensed, this share is significantly lower in medium and large companies (10.2% and 9.2% respectively). By contrast, research organizations (33.3% to 41.6%), universities (32.5%) and small companies (25.8%) license a much higher proportion of their patent portfolio. A recent survey carried out by the European Commission (EC, 2012b) finds similar results regarding innovative SMEs: they license on average about 23% of their patents.

Statistics on the trade of US patents yield similar results (Table 1.2). The fraction of patents filed by individuals or SMEs that are subsequently transferred ranges between 9% and 12%, while it falls to 6% for large companies (Serrano, 2010). A likely explanation lies in the patent owners' different abilities to exploit their patents internally. Large companies have usually a wide enough range of activities to find internal applications for many of their inventions. By contrast, smaller companies are more specialized, and thus more dependent on external partners for applications that do not correspond to their core activity.

1.3.3 Potential for enhanced patent trading

The PatVal survey shows that holders of European patents fail to license about 38% of the patents that they are willing to license, thus suggesting that – due to high transactions costs – the volume of activity in the market for patents is way below its potential (Gambardella *et al.*, 2007). Table 1.3 further highlights the distribution of patents that are either available for licensing or licensed by category of owner. It shows in particular that universities and small companies fail to license a significantly higher share of their patent portfolio (respectively 34% and 26%) than medium and large companies (9%). A survey of European innovative SMEs in turn suggests that respectively about 60% of their attempts to acquire patented technology, and 25% of their attempts to sell patented technology have failed (EC, 2012b).

The large share (36.3%) of unused patents in the PatVal survey (Table 1.1) might suggest an even stronger potential for enhancing the trade of patents. However, it is unlikely that all unused patents are actually suitable for sale or licensing. About half of them are kept for a 'blocking' purpose by their owner, while others might cover immature technologies and be kept as options for future uses. Moreover the distribution of patent value is skewed (Pakes, 1986; Pakes & Schankerman, 1984). A recent survey (EC, 2012a) of European patents granted to large European companies finds for instance that only 50% have a commercial potential, and that of those 40% have no value. Similarly, the top 3.5% of patents in the PatVal sample represent 77% of all patent value, while half of all patents account for little more than 1% of the total patent value in the sample.

Against this background, the potential for developing patent markets is probably limited to those patents that can generate large enough

Table 1.2 *Rate of patent trade in the USA by category of economic agent*

| | Individuals | | Companies | | | Public | |
	Not identified as professionals	Professionals	Small	Medium	Large	organizations	Total
Total (1)	304,087	17,654	453,683	567,081	565,582	25,383	1,933,470
Exchanged (2)	65.6	10.2	24.2	53,	31,540	809	170,470
Not exchanged	55.8	25.8	18.4	513,722	534,042	24,574	1,763,000
Ratio (2)/(1)	9%	12%	12%	9%	6%	3%	9%

Source: Serrano (2010).

Table 1.3 *Share of patents that are licensed or available for licensing*

	Willingness to license but not licensed	Licensed
Large companies	7%	9%
Medium companies	7%	9%
Small companies	13%	26%
Universities and other public research organizations	14%	34%
All categories	9%	14%

Source: PatVal (2005).

gains from trade to recoup the transaction costs. This is what a recent study of US patents sales initially held by SMEs also suggests (Serrano, 2011). It firstly shows that a sharp (50%) decrease of transaction costs would increase the average probability that a patent is traded from 23.1% to 29.6%. However this increase would in fact concern only a limited number of patents – among the most valuable ones – for which the probability of being traded increases by ten percentage points.

1.4 From casual transactions to organized markets: the growing role of intermediaries

Patent markets have always been heavily intermediated by professionals. The market for technology has developed in the nineteenth century in the USA thanks to both the change in patent laws and the proliferation of specialized intermediaries, lawyers and consultants in IP (Lamoreaux & Sokoloff, 2002). The acquisition of knowledge of the technology and the market is needed to reduce information asymmetries and the uncertainty associated with innovative projects: it assumes a learning process and high fixed costs, sources of economies of scale which can better be amortized over many operations (Hoppe & Ozdenoren, 2005). Hence the need for specialized entities, which can accumulate the required scale and experience.

Intermediaries facilitate the contact between providers and end-users for goods and services subject to information asymmetries and for which there is a significant cultural or geographical distance between participants in the exchange. They may also intervene when buyers and sellers prefer, for strategic reasons, to remain anonymous as late as possible in a negotiation. They can help during the negotiation and possibly act as counterparties to the exchange. To summarize the role of intermediaries:

- They produce information: they find partners (matching supply and demand).
- They bring expertise: being experienced in this type of transactions, they can help write the contracts and do the pricing.
- They can mitigate royalty stacking: if several patents with different owners cover the same product, the fact that individual owners seek to maximize their own income may result in an excessive price (the sum of prices of individual patents) and hamper demand. Instead, an entity that controls many patents involved in a particular technology field sets its price to maximize the total surplus produced, and not that of each individual patent (see also Chapter 5).

Below are outlines of representative business models of current IP intermediaries that may contribute to promote the smooth distribution of patents.

1.4.1 Intellectual property brokers and technology transfer offices

Intellectual property brokers and university technology transfer offices (TTOs) are mandated by their clients to support them in their IP sale or acquisition strategy. In that sense they are relatively close to a traditional model of intermediary in the patent market.

IP brokers

These IP-centric firms (e.g. IPotential, Inflexion Point, Thinkfire, Pluritas, ActiveLinks and Global Technology Transfer Group) provide technical, legal and business expertise to connect willing sellers and prospective buyers of patents and to complete patent transactions. By using such services, companies can expand their business opportunities, but also solidify their patent portfolios and thereby minimize the likelihood that an infringement claim will be made.

Intellectual property brokers operate on both the buy side and the sell side of patent transactions. On the sell side, they help their clients raise cash through the divestiture of some or all of their patent portfolios. Basically, they evaluate their clients' patents, select those that have some value and seek to identify potential buyers, based on their own connections and knowledge of IP marketplaces. In collaboration with clients, they develop a price target and sales strategy, and facilitate contacts and negotiations with potential buyers. On the buy side, IP brokers start by identifying acquisition candidate patents covering key technologies that are important for their clients' businesses. Then they approach owners of target patents and establish discussions regarding a potential patent acquisition or a licence for those patents, while preserving the anonymity of clients.

Technology transfer offices

Transferring technologies created by universities and research institutes to the commercial market is important to promote innovation. Entities like university TTOs have played a core role in facilitating technology and IP transfer/licensing from universities/research institutions to industry. For example, Stanford University Office of Technology Licensing (OTL) closed 107 new licence agreements, and received $62.5 million in royalty revenue from 546 technologies in fiscal year 2008 (Stanford University OTL, 2008). While some IP specialist firms such as IP funds and IP aggregation/licensing firms have recently begun to put energy into acquiring university/research institutes technologies, the role of TTOs as a platform to facilitate knowledge transfer from universities/research institutions to industry will become more important.

1.4.2 Trading platforms

Exchange platforms for IPR are pure intermediaries that facilitate the market matching process between IP sellers and acquirers. Unlike brokers or TTOs, they do not leverage their expertise to initiate and achieve over-the-counter transactions, but rather to encourage third parties to carry out such transactions by providing appropriate market infrastructures.

Internet platforms

Companies such as InnoCentive, Yet2, Tynax, UTEK, NineSigma, YourEncore, Innovation Exchange, Activelinks and SparkIP aim to facilitate the matching process between potential sellers and acquirers by centralizing information on available patents and/or buyer needs on web-based repositories. Besides web-based platforms, most of them offer a large range of side services to buyers and sellers, such as assessment of IP portfolios, support in formatting offers or assistance in negotiations. They are thus potentially interesting instruments to reduce the screening costs which, according to available evidence, are the most important factor of transaction costs in patent valorisation.

Intellectual property auction and intellectual property licence-right trading market

In contrast with mere web-platforms, the services offered by these intermediaries include price-setting mechanisms for the traded IP (see also Chapters 5 and 9). By remedying information asymmetries regarding the value of patents, they can thus further enhance transparency and predictability of the IPR market. Such IP transaction services include IP auction and IP licence-right trading market.

An IP specialist firm named Ocean Tomo holds live auctions for patents with the purpose of creating a highly transparent marketplace which facilitates the exchange of patents. Some entities – including Ocean Tomo, IP Auctions and Free Patent Auction – also provide online patent auctioning services. Their web platforms allow willing sellers to list innovative ideas protected by patents that are available for sale or license, and willing buyers to check if valuable patents are marketed. The creation of IP licence-right trading markets is even more ambitious in seeking to improve the transparency and predictability of IPR markets. Intellectual Property Exchange International is for instance planning to provide a highly transparent IP licence-right trading market called 'Unit License Right contract market'. If successful enough, such services may provide a price discovery function by allowing all market participants to monitor deal prices of individual patent transactions which have been confidential.

Patent pools

A patent pool is an agreement between two or more patent owners to jointly license one or more of their patents that are essential for a given technology as a single package (see also Chapter 5). It can also be defined as the aggregation of IPR which are licensed by patent owners to licensees through some medium, such as a specialized company or a joint venture set up specially to administer the patent pool (Clark *et al.*, 2000). Generally the administrator is authorized to issue non-exclusive sublicences to the patents on behalf of patent owners. The licensing revenue generated by the pool is then distributed among the patent owners according to a pre-arranged distribution method. So far most existing patent pools have been created to license patents that are essential to ICT standards (e.g. DVD, MPEG, MP3, Blu-Ray). Since the technical specifications of such standards are published in details, pools do not transfer additional knowledge to their licensees. Rather they license them the right to legally implement the standard.

As one-stop shops for licensing different patents related to the same technology, pools benefit to both IP licensors and licensees. They significantly reduce transactions costs by certifying (through independent experts) that licensed patents are valid and do really cover the technology, and by offering the same standard license to all licensees. They also facilitate the adoption of the technology by setting a unique price for the bundle of licenses, below the royalty stack that would have resulted had each owner offered a separate license. Accordingly, they provide an opportunity for patent owners to expand the market for their products by encouraging the spread of patented technology, especially in industries such as telecommunications, computer hardware and software, where companies have to access dozens or hundreds of patents to produce just one commercial product.

1.4.3 Patent funds

Patent funds are entities that invest in the acquisition of titles to patents from third parties, in view of achieving a return by monetizing these patents through their sale, licensing or litigation. This activity is akin to financial arbitrage: it primarily builds on the ability to screen the large population of patents in order to detect those that have an undervalued potential. In this respect patent funds can be expected to reduce screening costs. As intermediaries dedicated to patent valorization, patent funds are also potentially interesting instruments to reduce other transaction costs through specialization and economies of scale.

There is a large variety of patent funds in practice. A first key difference is whether they focus their activity on patent valorization through short-term enforcement of patent rights (or the prevention thereof) or through longer-term technology commercialization. The latter type of funds can in turn be differentiated according to their degree of involvement in the technology maturation process prior to commercialization.

A particular category of funds, known as 'patent trolls', refers to entities that exploit patents as liability rights only, for offensive purposes. They mostly operate in fields such as ICT, where the intensity of patent filings is such that a company must 'hack its way through a dense web of overlapping intellectual property rights in order to actually commercialize new technology' (Shapiro, 2001). They often use patents of dubious quality (of which many are actually revoked when the cases go to court).

Their approach consists in acquiring patents often from small patent holders or bankrupt companies in view of licensing or asserting them against companies that bring actual products or technology to the market. It features non-practising entities such as Acacia Technologies, Rembrandt IP Management or Constellation. Intellectual Ventures, a large patent fund, has also been undertaking this type of valorization strategy recently. These entities have been operating mostly in the USA, where the patent litigation system makes it easier to obtain high infringement damages or a profitable settlement. Since targeted licensees are companies that already infringe the patents, their activity does not bring about any development or actual transfer of technology. A famous case involves the fund NTP, which obtained a $600 million settlement from RIM (Blackberry) using five patents which were all revoked later. The threat used by NTP was that the Blackberry network might be shut down by the judge before the patents would perhaps be revoked, which would have ruined the business.

Defensive patent funds

Defensive patent funds have been created to protect operating companies from such aggressive forms of enforcement of patent rights. Their activity consists in preventively acquiring patents that could represent a threat to their clients. These funds may be for-profit entities such as RPX, or the ad hoc instrument of a group of large companies, as illustrated by the Open Innovation Network that covers Linux software. Recently, the government of Korea has created a public defensive fund to protect its national industry champions, and similar initiatives are reportedly ongoing in Taiwan and China. Besides acquiring patents, these national funds are meant to use their patent portfolio as a weapon to counter-attack against third parties that may sue their clients for infringement.

Box 1.2 Intellectual Ventures

Intellectual Ventures is a $5 billion (€3.5 billion) patent fund, whose business model consists in aggregating large patent portfolios and licensing them to third parties. Created in 2000, it is headquartered in the USA and has subsidiaries in Australia, Canada, China, India, Ireland, Japan, Korea and Singapore.

As of June 2009, Intellectual Ventures was reported to possess about 27,000 patents in various technology areas. It has created three funds, of which the largest one (Investment Fund I & II) is dedicated to investing in existing inventions acquired principally from individual inventors and SMEs. The other two funds are a seed capital fund (Invention Science Fund) and a fund that aims at extending the IP monetization system globally (Invention Development Fund I). In a presentation made in 2009, the President of Intellectual Ventures India, M. Suvarna, reported that there were 250 staff in Investment Fund I & II, versus 131 and 70 respectively in the other two funds. Although part of Intellectual Ventures' patents stem from R&D carried out internally or by its research partners (universities and research institutions), most of them have thus been purchased.

Intellectual Ventures targets companies for selling various licensing packages on a non-exclusive basis. Although its claimed objective is to invest in promising ideas and bring them to the market, Intellectual Ventures also seems to seek short-term valorization of patents as liability rights. Besides financial investors that have bought equity stakes, it is open to strategic investors (large ICT companies) that are simultaneously equity stakeholders and licensees of parts of its patent portfolio. In that respect, it is closer to a defensive fund. Moreover Intellectual Ventures has sold patents to other non-practising entities with more offensive monetization strategies (Wang, 2010) and recently started directly litigating potential infringers.

Patent trading funds and royalty funds

Other types of funds raise money from private investors to invest in the acquisition of titles to patents that will be licensed or sold to third parties. Acquired patents are aggregated into consistent sets, and transferred into Special Purpose Vehicles that handle their maintenance and commercialization within a given time period. Inventors receive an upfront payment for the patent, and in some cases a share of the royalty proceeds. Intellectual Ventures, a $5 billion American patent fund, whose business model consists in aggregating large patent portfolios and licensing them to third parties, is the most prominent example of this type (see Box 1.2). In Europe, Alpha Funds 1–3, Patent Select I and II and France Brevets are recent initiatives of this type. The Alpha Funds 1–3 and Patent Select I and II associate financial investors and IP consultants into a closed-fund structure. IP Bewertungs, the IP consultant involved in Patent Select I and II, filed for bankruptcy in summer 2011, suggesting economic failure in at least one case.

Royalty funds aim at acquiring titles to patents that are already licensed to a third party. Such funds provide capital to owners of royalty-producing IPR in exchange for their royalty revenues. They do not undertake any valorization activity, and their value added lies in their greater ability to hedge the risk associated with the royalty flows. Examples of this model include Royalty Pharma, DRI Capital or Cowen Healthcare Royalty Partners, all of which operate in the life science industry. While patent trading funds may fail to find buyers within scheduled term, and thus to reach the expected returns on investment, royalty funds are less exposed to commercial risks because the patents they target already generate cash flows before acquisition.

Technology development funds
Technology development funds aim to bridge the gap between invention and the exploitation of IPR when the technology is at a development stage too far upstream for its market applications to have the degree of certainty that other capital providers require. These funds have a comprehensive approach to IP assets, including not only patents but also the related know-how and equipment. Rather than purchasing 'naked' patents and nurturing them 'above ground', they typically invest directly in targeted companies, or in ad hoc corporate vehicles wherein the targeted IP owner transfers its R&D project and IP in order to isolate the transferred IP from the risk of the original IP owner's bankruptcy. Intellectual Ventures (through some of its affiliate funds), the German funds Patentpool Trust I and II and the Japanese Kyushu Investment Fund are examples of such entities.

Public authorities have created similar funds through public–private partnerships for public policy purposes. The German High Tech Gründerfonds or the European Investment Fund, a subsidiary of the European Investment Bank, are dedicated to early stage investment into projects or start-up companies, at proof of concept, pre-seed, seed, post-seed. In Italy, the Innovation National Fund was set up in 2010 to fund innovative projects carried out by SMEs to produce innovative goods based on patents and design. In Asia, the Innovation Network Committee of Japan (INCJ) has been set up by the Japanese government to enable large domestic companies to access inventions from public research organizations, start-ups and other companies. The various types of IP specialist firms are shown in Table 1.4.

Table 1.4 *Functions and business models of IP specialist firms*

Function	Business model	Specialist companies
IP brokers	Patent licence/transfer brokerage	Fairfield Resources; Fluid Innovation General Patent; ipCapital Group; IPVALUE; TPL; Iceberg; Inflexion Point; IPotential; Ocean Tomo; PCT Capital; Pluritas; Semiconductor Insights; ThinkFire; Tynax; Patent Solutions; Global Technology Transfer Group; Lambert & Lambert; TAEUS; etc.
TTOs	University technology transfer	Flintbox; Stanford University Office of Technology Licensing; MIT Technology Licensing Office; Caltech Office of Technology Transfer; etc.
IP trading platforms	Online IP marketplace	InnoCentive; NineSigma; Novience; Open-IP.org; Tynax; Yet2.com; UTEK; YourEncore; Activelinks; TAEUS; Techquisition LLC; Flintbox; First Principals Inc.; MVS Solutions; Patents.com; SparkIP; Concepts Community; Mayo Clinic Technology; Idea Trade Network; Innovation Exchange; etc.
	• IP live auction/Online IP auction • IP licence-right trading market	Ocean Tomo (Live auction, Patent Bid/Ask); FreePatentAuction.com; IPAuctions.com; TIPA; Intellectual Property Exchange International; etc.
	Patent pool administration	MPEG LA; Via Licensing Corporation; SISVEL; Open Patent Alliance; 3G Licensing; ULDAGE; One-Blue; etc.

Table 1.4 (*cont.*)

Function	Business model	Specialist companies
Patent funds	Short-term offensive	Acacia Technologies; Fergason Patent Prop.; Lemelson Foundation; Rembrandt IP Management; Intellectual Ventures; etc.
	• Defensive patent aggregation funds and alliances	Open Invention Network; Allied Security Trust; RPX; Eco-Patent Commons Project;
	• Initiative for free sharing of pledged patents	Patent Commons Project for open source software; etc.
	Patent trading funds	Intellectual Ventures; Alpha Funds 1–3; Patent Select I and II; France Brevets
	Royalty funds	Royalty Pharma, DRI Capital; Cowen Healthcare Royalty Partners
	Technology development funds	Intellectual Ventures; Patentpool Trust I and II; Kyushu Investment Fund; High Tech Gründerfonds; European Investment Fund; Innovation Network Committee of Japan; Italian Innovation National Fund

1.5 Conclusion

Like most institutional innovations, patent markets are a doubled-edged sword. They can be used for enhancing the circulation of knowledge using powerful market mechanisms, and they can be used as multipliers for abusive patent-based litigation. Reinforcing the positive, value-creation side will require the emergence and strengthening of a number of complementary institutions, both public and private. The quality of patents granted needs to be ensured so that no dubious title could be used for blocking innovative businesses. New types of intermediaries should strengthen the market infrastructure, facilitating matching of supply and demand, negotiation of contracts, evaluation of prices, etc.

The respective roles for governments and markets in this process will certainly vary across countries: whereas in the USA business actors are strong but the regulatory framework is weak, it seems that in many other countries, including in Europe, the reverse holds, pointing to the factors that currently block the development of patent markets and which need to be removed. The fact that patent markets are essentially global in scope might impose a process of convergence which will constrain national trajectories.

References

Anand, B. N. and Khanna, T. (2000) The structure of licensing contracts. *Journal of Industrial Economics*, **48**, 103–135.

Arora, A., Fosfuri, A. and Gambardella, A. (2001) *Markets for Technology: The Economics of Innovation and Corporate Strategy*, Cambridge, MA: The MIT Press.

Bessen, J., Ford, J. and Meurer, M. (2011) *The Private and Social Costs of Patent Trolls*, Law and Economics Research Paper 11–45. Boston, MA: Boston University School of Law.

Clark, J., Piccolo, J., Stanton, B. and Tyson, K. (2000) *Patent Pools: A Solution to the Problem of Access in Biotechnology Patents?* Alexandria, VA: US Patent and Trademark Office.

Cohen W. M., Nelson R. R. and Walsh J. P. (2000) *Protecting their Intellectual Assets: Appropriability Conditions and Why U.S. Manufacturing Firms Patent (or Not)*, NBER Working Paper 7552. Cambridge, MA: National Bureau of Economic Research.

European Commission (2012a) *Creating a Financial Market for IPR in Europe*, Final Report. Brussels: European Commission.

(2012b) *Options for an EU Instrument for Patent Valorization*, Report of the Expert Group set up by the European Commission. Brussels: European Commission.

Gambardella, A., Giuri, P. and Luzzi, A. (2007) The market for patents in Europe. *Research Policy*, **36**, 1163–1183.

Guellec, D. and Yanagizawa, T. (2009) *The Emerging Patent Marketplace*. Paris: OECD.

Guellec, D., Madiès, T. and Prager, J.-C. 2010 *Les marchés de brevets dans l'économie de la connaissance*, Report of the Conseil d'Analyse Economique. Paris: CAE.

Hoppe, H. C. and Ozdenoren, O. (2005) Intermediation in innovation. *International Journal of Industrial Organization*, **23**, 483–503.

Kamiyama, S., Martinez, C. and Sheehan, J. (2006) *Valuation and Exploitation of Intellectual Property*, STI Working Paper 2006/5. Paris: OECD.

Lamoreaux, N. R. and Sokoloff, K. L. (2002) *Intermediaries in the US Market for Technology, 1870–1920*, NBER Working Paper 9017. Cambridge, MA: National Bureau of Economic Research.

Ménière, Y., Dechezleprêtre, A. and Delcamp, H. (2012) *Le marché des brevets en France, 1997–2009*, Report of a study for Institut National de la Propriété Industrielle. Courbevoie, France: INPI.

Monk, A. H. B. (2009) The emerging market for intellectual property: drivers, restrainers and implications. *Journal of Economic Geography*, 9, 469–491.

Motohashi, K. (2008) Licensing or not licensing? An empirical analysis of the strategic use of patents by Japanese firms. *Research Policy*, 37, 1548–1555.

Nagaoka, S. and Kwon, K. U. (2006) The incidence of cross-licensing: a theory and new evidence on the firm and contract level determinants. *Research Policy*, 35, 1347–1361.

Pakes, A. (1986) Patents as options: some estimates of the value of holding European patent stocks. *Econometrica*, 54, 755–784.

Pakes, A. and Schankerman, M. (1979) *The Rate of Obsolescence of Knowledge, Research Gestation Lags, and the Private Return to Research Resources*, NBER Working Paper 346. Cambridge, MA: National Bureau of Economic Research.

PatVal (2005) *The Value of the European Patents: Evidence from a Survey of European Inventors*, Final report of the PatVal-EU Project, DG Science & Technology, Contract no. HPV2-CT-2001–00013. Brussels: European Commission.

Razgaitis, S. (2004) US/Canadian licensing in 2003: survey results. *Journal of the Licensing Executive Society*, 34, 139–151.

Serrano, C. J. (2006) *The market for intellectual property: evidence from the transfer of patents*. PhD thesis, University of Minnesota.

(2010) The dynamics of the transfer and renewal of patents. *The RAND Journal of Economics*, 41, 686–708.

(2011) *Estimating the Gains from Trade in the Market for Innovation: Evidence from the Transfer of Patents*, NBER Working Paper 17304. Cambridge, MA: National Bureau of Economic Research.

Shapiro, C. (2001) 'Navigating the patent thicket: cross licenses, patent pools, and standard-setting'. In Jaffe, A. B., Lerner, J. and Stern, S. (eds.) *Innovation Policy and the Economy*, vol. 1, pp. 119–150. Cambridge, MA: MIT Press.

Stanford University OTL (2008) *What Is Value?* Annual Report of Stanford University OTL. Stanford, CA: SUOTL.

Wang, A. W. (2010) The rise of patent intermediaries. *Berkeley Technology Law Journal*, 25, 159–200.

Zuniga, M.-P. and Guellec, D. (2009) *Who Licenses Out Patents and Why? Lessons from a Business Survey* STI Working Paper 2009/5. Paris: OECD.

2 | Strategic intelligence on patents

FRÉDÉRIC CAILLAUD AND YANN MÉNIÈRE

2.1 Introduction

All R&D managers face the same problem: they need to meet techno-logical challenges with solutions their researchers generally claim to be innovative and original. Yet ultimately, they may realize that these solutions are not novel, and cannot be used without buying a licence from the owner of the relevant patent. Indeed, most researchers in industry have a restricted perception of the field in which they operate, while the solutions they are looking for may already exist in another sector (19 million researchers publish about 18,000 articles every day, and there are about 60 million patents). Against this background, the ability to effectively screen and analyse the innovation landscape represents a major stake for companies.

Patents are the earliest markers of innovation: most of the information they contain is not available elsewhere. Therefore it is not surprising that firms rank them as one of the best information channels on their rivals' R&D (Cohen *et al.*, 2002). Accessing patent information has actually become critical in the current 'open innovation' era, where-by firms increasingly rely on third parties to acquire or commercialize technology. Companies cannot afford to adopt a development strategy without analysing as a first step the global patent environment, in order to optimize their R&D productivity, select the most suitable partners and/or anticipate their competitors' strategic R&D orientations.

Nevertheless, drawing critical information from the patent literature is far from being an easy task. Patent searches only provide imperfect signals that must be interpreted carefully before making a strategic decision. Even extensive searches by skilled professionals do not guaran-tee that all relevant patents can be identified. Moreover, the description of

We are grateful to Anne Mazzucotelli, Jean-Yves Legendre (both from the Licensing Department at L'Oréal) and Valérie Mérindol (Observatoire des Sciences et Techniques) for allowing us to use the pictures that illustrate this chapter.

the protected invention disclosed in the patent seldom provides a full picture of that invention (Hall & Harhoff, 2011). This makes it particularly difficult to identify patented inventions that present a real economic potential. The legal strength of a patent – that is its effective scope and validity – is another factor of uncertainty; it must be assessed before deciding, for instance, whether to start a project or purchase a licence. Addressing all these limitations requires advanced legal, scientific and business skills. Accordingly, the process of finding, identifying and assessing potentially relevant patents is long and costly, with uncertain results.

The steep increase in patent applications that has taken place in recent decades has made the task even more difficult. Due to the larger number of patents, it is at the same time more likely that any technological solution infringes prior patents, and more difficult to identify each of them. The patent surge has also accentuated the heterogeneity in patent quality. A survey carried out in Europe concludes for instance that about one-third of European patents granted are not exploited, either because they are used as weapons to block competitors, or because the underlying technology is not exploitable in the market (Gambardella *et al.*, 2007). Frequently, the latter patents are also legally fragile. As a result, information asymmetries undermine firms' ability to exploit patent information effectively at the very time when their need for such information is becoming critical.

Over the last years, this tension has led a number of major players to use sophisticated statistical instruments to supplement the traditional labour-intensive approach to patents. Some recent analytical and visualization methodologies indeed enable a much deeper exploitation of the wealth of information available in patent databases, with a wide variety of applications (from internal portfolio management to strategic intelligence on rivals and partners). By casting new perspectives on the moving landscape of innovation at an unprecedented scale, these instruments currently confer a key strategic advantage to their users. In the long term, their democratization can be expected to have an even deeper impact on the whole innovation ecosystem, by significantly enhancing its transparency for all stakeholders.

This chapter aims to present the methodological principles underpinning these instruments, and to explain how they can be used in practice to inform strategic decisions. Using practical examples, we describe in turn the two main categories of instruments currently available, namely visualization methodologies (Section 2.2) and patent portfolio quality analyses (Section 2.3). We review possible applications of each type of instrument, and highlight their advantages and limitations, and the current barriers that still impede their use by a larger number of actors. As a conclusion to the chapter, we seek to derive the medium- and long-term implications of the ongoing process of improvement and diffusion of these instruments for the whole innovation system.

2.2 Recent analytical and visualization methodologies enable a much deeper exploitation of patent information

Traditionally, patent analyses carried out as part of freedom to operate studies have been done manually and 'microscopically': the patents are first identified by means of bibliographical searches and are then analysed line by line by specialist engineers. Given the skills and time required to process each patent, only a few dozen patents can be analysed, at the risk of neglecting opportunities that might be found in their immediate environment.

For some years now, the advent of new specialist patent portfolio management software has been extending these limits. The most recent versions of automated patent database processing software (Questel's Orbit and soon Thomson Innovations by Thomson Reuters) can for example analyse up to a million patents simultaneously. Aside from the considerable productivity gains, they are literally opening up new horizons to analysts, bringing about a complete change in the way companies develop their innovation strategies.

In this section, we provide an overview of these new possibilities, based on the example of the Thomson Reuters software (which has the advantage of producing very intuitive results). This software has been around for over 10 years, but it only became truly successful at the end of 2009 when it partnered up with the Derwent patent database, one of the most extensive databases available. Before presenting its applications, we briefly explain the way it functions.

2.2.1 *Mapping by topographic visualization*

The Thomson Innovation software (from Thomson Reuters) provides a **navigation tool for innovation**, based on the representation of vast technological fields in the form of topographical maps, where the relief shows patent density.

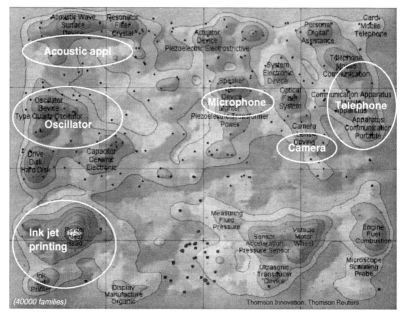

Figure 2.1 Detection of cosmetic innovations in the piezoelectric sector. See plate section for colour version.

Each map can contain up to 50,000 patents, soon to rise to 500,000. The patents are divided across the map according to statistical assessments of the occurrence and recurrence of the words used in their title, summary and claims. Each word is attributed a weight based on its frequency in the corpus and the inverse document frequency. Rare or too-frequent terms are eliminated to create a set of *n* relevant descriptors (the summaries are also rewritten with identical defined semantics, allowing for more efficient identification of the descriptors). A naïve Bayes classification is applied. Each patent is then linked with a set of one to four vectors in an *n*-dimension space defined by these descriptors, and is therefore positioned close to similar patents. This space is then translated into a self-organizing two-dimensional topological map. To facilitate visualization of the results, topographical contour lines and colour shading transcribe the innovation density on the map (Figure 2.1).

This tool's appeal lies in its currently unrivalled visualization of the information contained in an extensive corpus of patents. In practice, it displays mountainous islands in the middle of the sea. The higher the patent mountain, the denser the innovation in this field. The closer one mountain or hill is to another, the more similar the subjects dealt with in the patents. By reading the key words on each mountain we can determine the patents' scope of application. Just like Google Earth, we can zoom in on regions of interest, from an entire sector right down to the individual researcher.

If one performs an annual analysis of the patents filed in a given sector, one can detect innovations through the emergence of an island (rapid increase in the number of patents located in the same area of the sea). The objective of the map in Figure 2.1, for example, is to detect cosmetics applications from an analysis of the piezoelectric sector.

These two sectors are supposed to have nothing in common, but not only is an island forming (patents in red), but one of the patents has already led to a product launch in the cosmetics field.

2.2.2 Competitive patent environment: impact on strategic decisions

The technical prowess that constitutes the mapping of a technological field provides innovation players with a revolutionary tool for grasping and interpreting their strategic environment. It can be used to display a very clear overview of all the players present (competitors, future competitors or potential partners) together with their positioning; it can even analyse the dynamics between them. Below are several examples.

Map of operators within a field
The Thomson Innovation software's main application is the creation of 'Ordnance Survey' maps showing the positions of all those involved (subcontractors, suppliers, independent inventors, etc.) within a particular field of innovation by using different coloured dots. Figure 2.2 presents a map of the 25,000 patent families in force in the medical device sector. It was divided into zones corresponding to this sector's main applications. The snowy mountains indicate a very high density of innovation, and therefore mature application fields. The reverse is true in the sea (blue ocean), which represents breakthrough innovations.

The patents of the three main sector players are indicated by different coloured dots. The map therefore makes it possible to identify immediately the specialist companies whose patent portfolio gives them a dominant position in a specific application – like Medtronic

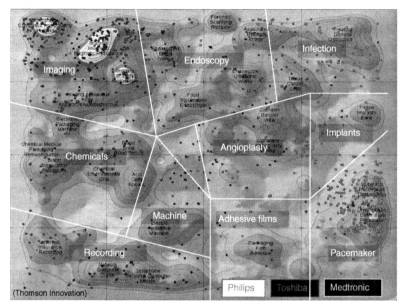

Figure 2.2 Patent mapping for medical devices. See plate section for colour version.

for pacemakers. It also identifies sectors facing opposition (different coloured patents very close to each other, like those of Toshiba and Philips in imaging). You simply click on a patent dot to access and read it.

Map of a country's technologies
The development of certain technologies is not divided evenly throughout the world. The map shown in Figure 2.3 has been established in collaboration with Observatoire des Sciences et Techniques. It compares the application strategy of European public research organisations in the field of recombinant DNA in France (grey), Germany (yellow) and Great Britain (red). A dominant position results from the filing by the public research organizations of the same country of a cluster (also called thicket) of patents in the same area of the map.

France does not really have a dominant field that it might have been able to appropriate by filing a 'minefield' of patents for strategic purposes, unlike Germany in the field of plants or Great Britain in the medical field. This map is particularly instructive for a country wanting to establish the dominance of their public research in a given sector. It identifies the patents it needs to buy to strengthen its portfolio

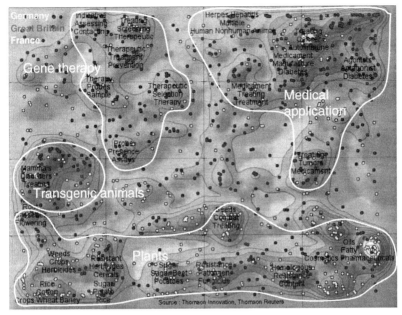

Figure 2.3 European public research organizations in recombinant DNA. See plate section for colour version.

and the institutions it should join forces with to become front-line players and launch a successful licensing-out programme. It can also identify any 'holes' left by the competition, indicating areas for additional research and where consolidating patents should be filed.

Analysis of competitors' R&D strategies

Figure 2.4 presents a dichroic analysis of the patent applications filed by two companies between 1995 and 2000 (blue dots), 2000 and 2005 (yellow dots) and 2005 and 2009 (red dots). The coloured circles (blue, yellow and red) are defining the fields of research at different periods of time for the two companies. It shows that the two companies have a similar R&D strategy but company A is clearly trying to be more innovative by exploring different and wider fields.

Selecting the most suitable partner

Figure 2.5 is an example of the way patent mapping can be used to search for potential partners for the acquisition and development of technology. In the field selected (20,000 patent families), the patent

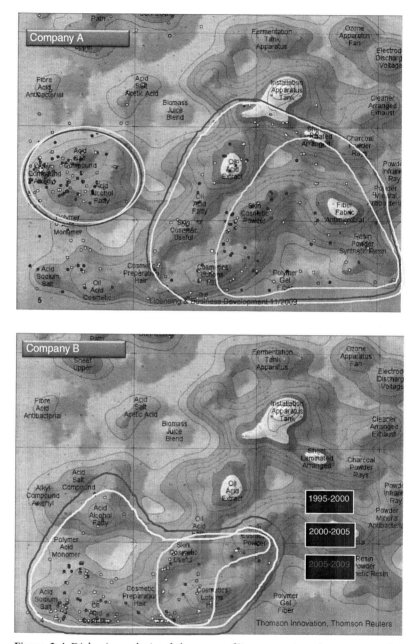

Figure 2.4 Dichroic analysis of the patent filing strategies of two companies from 1995 to 2009. See plate section for colour version.

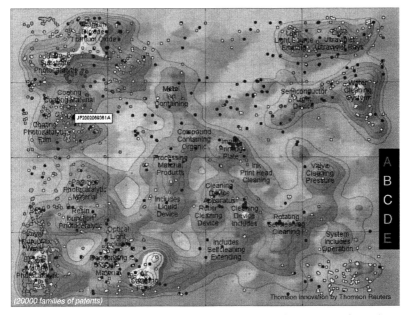

Figure 2.5 Selecting the most suitable partner. See plate section for colour version.

portfolios of two companies are indicated by different colour codes. If the objective of a desired partnership is well defined from the outset, the ideal partner will be the one with the biggest patent portfolio in the specific field. Companies A, B, C and D have established patent 'minefields' in very specific fields, making them the undisputed leaders. If the objective is not very precise or if several technical solutions could meet the same need, it is best to choose a partner like company E or C, with patents spread practically all over the map.

When the potential partner has been selected, the next task is to determine the potential complementarities. Here, it is useful to add up the patent portfolios of the two companies (maximum 50,000 families at present), allocating them different colour codes. This technique is called cross-mapping. No other method can compare two such large patent portfolios so quickly and efficiently and yet so intelligently. In the example in Figure 2.6, company A (patents in red) wanted to work with company B (patents in green).

We can see red patents in the green region and vice versa; these can be read by simply clicking on them. This means that each

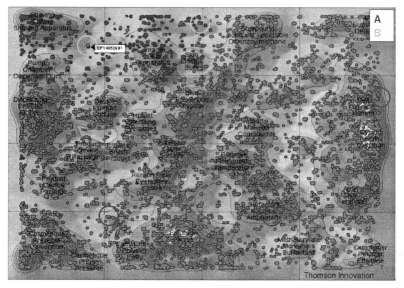

Figure 2.6 Cross-mapping. See plate section for colour version.

company has inventions that could be of interest to the other and become the subject of a partnership or licence. If one of these patents has led to a product launch, there may not be freedom to operate. In the circles, we can see several examples of patents in one colour surrounded by many patents in another colour. This method therefore detects areas for potential infringements and/or collaborations. This analysis can also lead to the set up of cross-licences. Cross-mapping must therefore be performed before any meeting with a potential partner.

2.2.3 Impact on the companies and public research organizations

Patent mapping is a potentially critical tool for all companies that conduct research, in a context where it is no longer possible to invest in a new idea before checking its freedom to operate. It allows the company to challenge new project holders before making any R&D investment, by asking them how the projects are new and differentiated. At the moment, the best way of answering this question is by creating a world map of the patent environment in the field of interest, and then positioning on this map what you hope to do. Straight away, it will indicate whether the theme is new (it will be located in the sea) or whether it is the subject of many patents (mountainous area). In the latter case, it would be useful to zoom in on the players present on this mountain, analyse their patent portfolios and determine which of them would be the best partner for the project. In an era of open innovation, this approach would seem essential.

However, only a very small number of companies are currently using these visualization tools, due both to their high access cost and a lack of know-how. Getting the best out of tools like the Thomson Reuters software requires real expertise, which is still limited to a small number of experts in the world. While access suppliers may offer training on the functionalities of these tools, they do not know how companies use them, and these companies are reluctant to communicate their methodology because they, quite rightly, consider it to be a major competitive advantage.

This software constitutes a particularly effective decision-making tool with regard to strategic analysis, albeit imperfect since mass data analysis always produces inadequate information. Therefore visualization mapping makes it easier to identify licensees and potential infringers, but it does not enable one to do away completely with the more traditional analyses prior to decision-making. The patents identified must be read and analysed by an expert, the only person truly able to interpret all of the information.

2.3 How is the quality of a patent portfolio analysed?

The main limitation of visualization tools lies in their inability to sort patents correctly according to their value. Simply identifying patents in an area of interest does not necessarily mean these patents are valuable, far from it! Their quantity is no guarantee of their quality. According to the PatVal (2005) survey, only about two-thirds of European patents are actually exploited – covering products and services already on the market or whose market potential is definable. The remaining third is made up of 'bare' patents that are part of a defensive strategy or patents for which it is still too early to imagine the market potential. Moreover, patent offices like the US Patents and Trademarks Office have long had a reputation for granting patents that should not have been in view of the available prior art. Finally, many patents have not yet been granted. In any event, it is advisable to analyse the quality of each patent identified as being of potential interest. Combining a quantitative mapping analysis with a qualitative analysis of the patents will therefore refine any necessary strategic decisions.

It is worth clarifying at this stage that the notion of classifying or rating a patent refers to a patent's intrinsic quality – i.e. the invention's technological merits, its potential market and the quality of the legal cover – and not its potential financial value (see Box 2.1). These two concepts of value are often confused when they are actually very different. Deducing the potential financial value of a patent from its qualitative classification remains an impossible task, even if a quality patent should, in theory, generate better profits.

2.3.1 The traditional method

The rating of patents is traditionally entrusted to a multidisciplinary team responsible for assessing (based on their experience) the technical, legal and commercial parameters that affect the quality of a patent depending on what the company plans to do with it.

In practice, the company gathers its evaluators together at regular intervals to specify its extension and valuation policy and define the patents to be abandoned. The number of parameters potentially influencing this rating is so great (over 100) that companies which have a sizeable diversified portfolio are generally happy to use simplified rating methods, because an in-depth analysis would be too long and too costly compared with the average potential value of a patent. This type of rating is highly subjective because it rests entirely on the quality of the experts involved and on their ability to analyse as impartially as possible patents they have very often filed themselves. A rating can also be suddenly challenged following the publication of new patents, and therefore needs to be updated as frequently as possible. We can easily see the difficulty faced by companies that have to evaluate a sizeable patent portfolio in just a few days following an acquisition.

Faced with this complexity, multiple sector operators regularly put forward 'new' rating methods to companies and public research organizations (PROs), although it is very often a question of variations on the same principle. Each parameter is attributed a grade, for example from 1 to 3. The grades of all of the parameters affecting the invention's potential market are added up and then weighted (often

---

Box 2.1 The financial valuation of patents

The potential financial value of used patents is established using recognized methods that have been used for decades (discounted cash flow, real and similar costs, etc.). It is interesting to note that no international standard has yet been introduced in this field, which often makes final negotiations tricky because it is very rare even for two experienced evaluators to arrive at similar orders of magnitude of value. The only country that has recently tried to create a standard is Germany, through its standardization institute, the Deutsches Institut für Normung (DIN). The national validation process is very advanced, and the DIN is currently trying to convince other European states to accept this standard. The Association Française de Normalisation (AFNOR) in France has also initiated a similar approach, without greater success for the time being. Despite clarification efforts, the proposed framework is still far too flexible to enable two independent evaluators to arrive at similar financial valuations. The sheer number of parameters to be taken into account makes it difficult to find a compromise that suits the patent valuation experts. We should welcome these approaches, but we are still a long way from obtaining a sufficiently instructive and precise international standard that will compensate for the current lack of consensus.

The only exception is the pharmaceutical/biotechnological sector, in which the financial conditions of most transactions are published and analysed, creating a financial scale of charges that is regularly updated according to factors like stage of development and field of application, which sellers and buyers refer to.

Conversely, the 35% of patents that are 'bare', which are part of a defensive strategy or whose market potential does not yet exist are especially difficult to evaluate. The scarcity of public data on the transactions of 'bare' patents and the absence of a commonly recommended method or approach are major obstacles to their valuation and represent a challenge for potential buyers and sellers.

---

arbitrarily) with the grades from the legal and technological analysis. The global average is then supposed to reflect the relative quality of a given patent in a portfolio. Obviously, because companies do not use standardized approaches, this rating can only be used in-house and it would take much stronger arguments to convince a buyer of a particular patent's importance.

Box 2.2 IPscore and similar systems

The European Patent Office has offered small and medium enterprises (SMEs) a tool for analysing the quality of their patent portfolio called IPscore. Patents are classified by answering a long list of questions about them. Each parameter analysed is graded from 1 to 5, and all of the data is put together in the form of radar charts. This rather exhaustive questionnaire becomes tedious, however, if you have more than ten patents to analyse. It is an interesting back-up tool, but it is too small-scale and subjective to analyse bigger patent portfolios or patents from unknown fields.

A large number of private firms have developed systems based on the same principle, but whose final objective is to accumulate the grades. Rating a parameter from 1 to 5 is acceptable; however, when using a dozen or more weighting them to give an overall grade is much more questionable.

While a combination of excessive simplifications and complex and poorly mastered rating systems might make the consultants happy, it does not sit well with the entities wishing to value the research they have financed (see Box 2.2 for an example of a portfolio analysis tool). The potential value of a patent is generally underestimated when using the traditional method because there is not enough information to be able to position this invention in its true global context. In an era of open innovation, statistical analysis and patent mapping tools, it is a loss of value that is difficult to come to terms with.

2.3.2 The statistical methods

The statistical analysis of patent databases is an interesting alternative to the more traditional approach historically used by companies. By adding up the information extracted from 60 million existing patents – including their respective patent approval procedures and litigation history – it is possible to have access to an enormous amount of information, which can be used to help better define the relative quality of a portfolio's patents. Then it is a case of sorting it, categorizing its impact and giving it meaning for the practitioners managing the patent departments. For 10 years, over 100 econometric evaluation studies have allowed to identify a set of indicators (non-exhaustive list provided in Table 2.1) more or less correlated to the quality of the

Table 2.1 *Main indicators of patent quality*

Number of independent claims
Number of dependent claims
Average length of independent claims
Shortest independent claims
Types of claims
Size of patent family
Patent class/subclass
Patent pendency period
Scope and content of cited prior art
Relative earliness of priority date
Forward citation rate
File history details
Number of related patents

patents. These predictive parameters shed a new, useful and objective light which is complementary to that obtained by the traditional method. The informative value of each of these indicators is very low, however, when they are considered individually. Their use as decision-making tools therefore requires a global approach, combining different relevant indicators according to the issue being addressed. The examples below illustrate this approach.

Generality and Originality Indices

The Generality Index is derived from the forward citations of a patent. It aims to capture the patent's influence on subsequent innovations, taking into account the variety of fields in which the patent is cited. The higher this index is, the greater the likelihood of finding licensees in various domains. The Originality Index is based on the calculation of backward citations (cited patents in a wide range of fields). The higher this index is, the further the invention's reach. By way of example, Figure 2.7 reports the value of these two indices for a set of patents from the same portfolio in a particular field of cosmetics. The top right box shows all the patents from this portfolio with the highest Generality Index and Originality Index, that is, the patents most suited to being evaluated. The bottom right box, on the other hand, corresponds to the patents that are the least well suited to being evaluated individually.

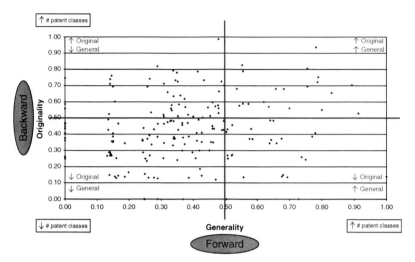

Figure 2.7 Generality and Originality Indices. See plate section for colour version.

The acceleration coefficient

The acceleration coefficient is used to compare hundreds of companies or PROs within a given sector, and to determine which of them have recently taken out lots of patents in a field over a given period of time (Figure 2.8). Each company is represented by a coloured circle. The diameter of each circle is directly linked to the sales of each entity in the field. The higher this index, the more the company has identified a promising invention. It is currently one of the best methods – along with mapping – of identifying a breakthrough innovation.

Company-specific patent signature

L'Oréal has created a database in which each of the products launched by its competitors is listed and linked to the patent that protects it. We have therefore been able to analyse the patent portfolios of several major cosmetics players to determine the time elapsed between the first patent application and the launch of the corresponding product(s). We have been able to demonstrate that the cumulative filing of patent applications is triggering a product launch if the number of patent applications is crossing a threshold which seems different from one company to another. This calculation allows one to anticipate a new product launch by the company studied, 3 to 5 years prior to its effective launch. Our analysis does not allow one to define accurately

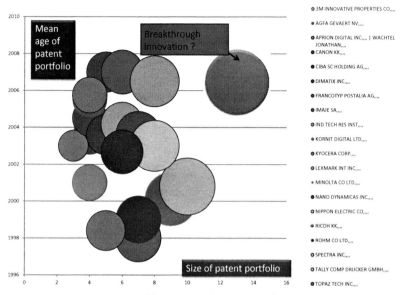

Figure 2.8 Acceleration coefficient. See plate section for colour version.

the profile of the product but, due to a personal communication, it is feasible. As the thresholds are different from one company to another, it could be considered as a signature tightly linked to the company R&D culture.

Cross-citations

It is also possible to retrace cross-citations between operators in the same sector, and compare their evolution over time by using Orbit software developed by Questel. When two operators are citing each other, they are linked by a straight line. The number in black indicates the number of backward citations, the one in red, the forward citations. While most of the operators are citing the leader, the diagram in Figure 2.9 shows the appearance of a series of cross-citations which no longer pass through the leader but between companies A, B and C. Since the sector leader had refused to grant licences on its technology to most of the operators in the field, companies A, B and C decided to join forces to develop radically different innovations to those that existed before. The cross-citations analysis allows the early identification of the emergence of clusters of operators that are collaborating together to develop products based on similar concepts. It is a useful tool to locate innovations more easily.

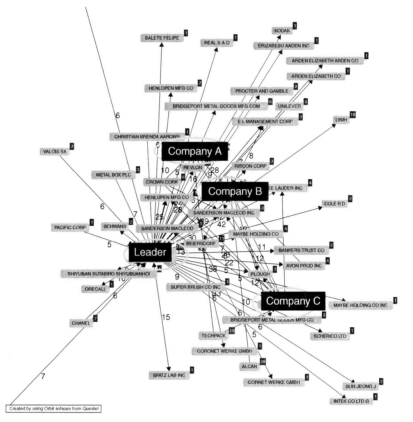

Figure 2.9 Cross-citations. See plate section for colour version. (Diagram created by using Orbit software from Questel.)

2.3.3 Automated tools

Econometric work on patent databases has led to the emergence of the first companies offering automated tools to assess and grade the quality of patents. Due to the importance of the American market for patent valuation, all these initiatives have first taken place in the United States, whereas Europe and the rest of the world are only starting to discover these instruments.

Several automated rating tools have been developed by, among others, the IPQ from Ocean Tomo. In almost all cases, the demonstration of the relevance of these algorithms is based on unpublished but patented work, which makes any attempt to evaluate their respective performances difficult. These methods patents (no description or justi-fication of the parameters or their weighting in the algorithm) have hindered the development of more relevant and transparent tools in the USA. And yet, it would not have been possible to elaborate the core design of these instruments without discussions and confrontations between their creators and econometricians. Similarly, the reasoning behind the rating should be as transparent (not a black box) and controlled as possible so that different independent evaluators using the same method can reach very similar results that can be understood and accepted by the community of practitioners.

At this stage, automated rating tools have therefore been relatively disappointing operationally, because they do not take into account the business model of the sector concerned, the different types of use of the patent (licences, litigation, defence, etc.) and the other geograph-ical areas (Germeraad, 2010). Moreover, giving a patent an overall accurate grade without indicating the confidence interval and without providing explanations as to how it was arrived at is unsatisfactory. More generally, an automated tool will probably never be satisfactory enough to replace the traditional evaluation method, based on privil-eged, unpublished information. It will, however, provide important complementary information that can help users make better docu-mented decisions and get the best out of the fruits of the research.

2.3.4 Impact of the patent rating potential on companies

The use of statistical rating methods will probably become much more widespread in the near future. Indeed, mastering this type of tool gives operators a major competitive advantage. Moreover, the most innova-tive companies can no longer accept being judged solely on a parameter as broad as the number of patents they have filed, given the growing number of intangibles involved in their valuation (up to 85% in many activity sectors). Statistical rating will soon enable any investor to analyse the portfolio of patents published by any player within a sector in order to compare their respective strategies and performances on valuation. Two companies that appear similar in terms of the number of patents filed can then turn out to be very different when the qualita-tive distribution of their respective patents is taken into account.

Nonetheless, the traditional method of evaluating patent quality is still heavily dominant at present. The number of patents filed is also still the most frequently used innovation indicator used in companies and PROs today, even though it is already possible to sort a patent portfolio according to well-established predictive parameters. Aside from the technical difficulties posed by the evaluation of the potential value of a 'bare' patent (i.e. a patent whose potential market cannot yet be defined, or whose purpose is purely defensive), this persistence of traditional patent portfolio management methods can largely be explained by a lack of information and training among the vast majority of practitioners. Indeed, the training received by managers and teams in patent departments remains purely legal, and focused on the approval and defence of patent applications. Company directors and financial analysts are also insufficiently informed, whereas they could serve as opinion leaders.

From this perspective, the key to changing current practices resides in training practitioners in statistical methods, to enable them to assess the relevance, but also the limitations of the different predictive parameters. At present, very few agents or patent consultants are capable of offering this type of service, and their cost is still beyond the reach of SMEs. There therefore remains a latent demand for more structured training by specialists or suppliers of patent databases. This could take the form of a theoretical and practical training course in the econometric analysis of patents, aiming to help companies and PROs adapt themselves to this evolving environment.

The emergence of effective, objective rating tools approved by the international community should also be a powerful catalyst, speeding up their integration in companies and PROs. These entities will then be able to better analyse and position the quality of their portfolio in relation to those of their competitors, but also to communicate their analyses to third parties. The existence of recognized rating tools could for instance give rise to information campaigns initiated by companies (in the annual report or in the press) aiming to explain how the patent portfolio is strategically aligned, how it compares qualitatively to the competition, and how it is managed and optimized from a valuation point of view (licence revenues, transfers, acquisitions, creation of clusters, etc.). It should also encourage the spontaneous supply of this information to financial analysts to establish the value of patent-based intangibles and help investors better define the technological future of the companies. Eventually, it would not be surprising to see company directors in the USA being questioned on their IP policy by experts appointed by major investment funds during annual general meetings. Indeed, new insurance is already being launched in the USA for patent department managers to protect them in the event of shareholder complaints.

2.4 Conclusion

The search for prior art in patent databases is a necessary yet difficult task for all firms and PROs engaged in R&D activities. During the last decades, the growing volume of patents has made this task even more challenging. At the same time, the transition towards the open innovation paradigm was further reinforcing innovators' needs for strategic intelligence about their rivals and potential partners' R&D activities. These evolutions put traditional methods for analysing patent information under strong pressure: the human resources and time they require are hardly compatible with the ever-increasing amount of information to be processed. Against this background, new methodologies enabling the statistical exploitation of patent data on a very large scale offer increasingly relevant solutions, thereby paving the way for a deep renewal of the way in which firms elaborate their R&D strategies.

We have shown in this chapter that such innovative instruments can produce critical strategic intelligence by enabling the visualization and quality assessment of entire patent portfolios at both the macro (country, sector) and micro (firm) levels. Besides searching for prior art, they make it possible to assess quickly the strengths and weaknesses of any firm's IP portfolio, anticipate rivals' strategic R&D orientations and identify infringers or potential partners for technology transfers. Accordingly, they confer significant strategic advantages to the few actors who can already afford to use them.

Of course these new instruments are not perfect, and they will certainly never suffice to fully replace human expertise in carrying out patent analyses. However, they will be further improved, and they are bound to reach a much larger public in the coming years. Their generalization is then likely to induce major evolutions in the innovation ecosystem, by introducing transparency in what still remains one of the more complex and opaque facets of the economy. Likely outcomes include a more efficient allocation of R&D investments, enhanced possibilities for financial investors to screen companies according to their innovative performances, but also the development of large markets for patents at the global scale. Despite their growing importance, such markets indeed remain characterized by strong information asymmetries that hamper the matchmaking process between supply and demand and sellers. In this respect, enhanced transparency can be a powerful leverage to alleviate transaction costs, and thereby encourage a more efficient allocation of disembodied technology through the trade of patents.

The transition towards this transparency paradigm can be facilitated by the development of a competitive supply for new instruments. It will also depend on the pace at which potential users become aware of them and acquire the technical skills required to exploit them. Meanwhile, the high price of these instruments and the lack of competences remain significant barriers to their adoption. The longer these barriers persist, the longer their use will remain restricted to a small number of entities – such as multinational companies or recently created patent funds – that can leverage this information edge to build particularly powerful patent positions for the years to come.

References

Cohen W., Goto A., Nagata A., Nelson R., Walsh J. (2002) R&D spillovers, patents and the incentives to innovate in Japan and the United States. *Research Policy* 31, 1349–67.

Gambardella A., Giuri P., Luzzi A. (2007) The market for patents in Europe. *Research Policy* 36, 1163–1183.

Germeraad P. (2010) It is all in the game. *Intellectual Asset Management* 39, 63–71.

Hall B., Harhoff D. (2011) *Recent Research on the Economics of Patents*, NBER Working Paper 17773. Cambridge, MA: National Bureau of Economic Research.

PatVal (2005) *The Value of European Patents: Evidence from a Survey of European Inventors.* Brussels: European Commission.

3 | Microeconomic foundations of patent markets: the role of intermediaries, auctions and centralized markets

ANNE PERROT AND
ANNE YVRANDE-BILLON

3.1 Introduction

The issue of 'markets for ideas' has already been examined by a series of theoretical or empirical contributions, investigating the forms these 'markets for knowledge' might take (Arora and Gambardella 2010; Gans and Stern 2010). Markets for knowledge are larger than patent markets, as patents are a precise formalization of ideas and knowledge, which, in principle, allow the identification of their scope and coverage. Some articles dealing with markets for knowledge examine less formalized knowledge exchanges, like the open-sourced Wikipedia encyclopedia (see e.g. Tapscott and Williams 2008): this encyclopedia is at the other extreme of the spectrum of ideas exchanges since its operation is based on the free availability of information or ideas provided by authors who do not look for profitability, but want knowledge to be spread as widely as possible.[1] In this chapter, we exclude such exchange modes from our analysis to focus on patents. Indeed, a priori, patents are items that are easier to identify and exchange than 'ideas', 'knowledge' or 'information' because their outline is more precise.

And indeed, over the last three decades, inter-firm transactions over patents have grown exponentially. Licensing agreements have multiplied (Athreye and Cantwell 2007; Mendi 2007), intermediated marketplaces have developed (Howells 2006) and patents auctions have emerged (Jarosz et al. 2010).

However, the introduction of market mechanisms to coordinate patents exchanges has not eradicated the other modes of knowledge exchange. Indeed, various institutional arrangements coexist for technology transfer, ranging from negotiation over the counter to auction through hybrid mechanisms, such as joint ventures or cross-licensing.

---

[1] We do not discuss, here, the quality of this knowledge.

Moreover, the development of markets for patents seems to be slow. While most of the companies surveyed on their strategy for patents and licences indicate a strong growth in revenues generated by these assets, the majority also highlights the inadequacy of their licensing activity relative to their expectations. Thus, according to Arora and Gambardella (2010), while 18% of the European Patent Office patents are offered for licensing, only 11% are actually licensed. Furthermore, large firms (i.e. with more than 250 employees) are willing to license 16% of their patents, but succeed in licensing only 9%, and small firms (i.e. with fewer than 100 employees) are willing to license 37% of their patents, but only license 26%. Similarly, using a survey of 9,000 European patents (the PatVal-EU Project), Gambardella *et al.* (2005) find that 36% of the patents are not deployed for internal use or licensing. Among these patents, nearly 52% (or 18.6% of all patents) have a strategic value for companies that hold them as they help block competitors, the remaining 48% (or 17.4% of total patents) being 'sleeping patents', without any strategic value, and which could be sources of value if they were marketed (see Guellec and Ménière, Chapter 1, for more details on this issue).

Finally, the few experiments of patent auctions that have recently been organized proved not to be very successful. According to available data on the results of the auctions held by the US company Ocean Tomo, over the 749 lots of patents auctioned since spring 2006, only 38% have been sold, and at less than 80% of the expected price on average (Jarosz *et al.* 2010).

These data, even if they provide a partial overview of reality, suggest that market transactions on such particular goods as patents are far from being systematic. Does it mean that patents do not easily lend themselves to market exchange? Or is it due to failures of the existing modes of market coordination? In other words, what are the institutional and structural obstacles to the development of markets for patents? The purpose of this chapter is to provide some answers to these questions by referring to recent developments in the literature on market design and transaction costs economics.

For that purpose, in the first part of the chapter, we review the theoretical properties of the various coordination mechanisms and show that the efficiency of the modes of exchange depends on the characteristics of the traded goods and on the information available to agents. In the second part, we describe the 'patents' goods by focusing on the characteristics that hinder their marketing. The third section discusses the potential for development of patent transactions as well as the possible or desirable role of policy makers.

3.2 Properties of the various exchange mechanisms

3.2.1 Markets and hybrid forms

Transaction cost economics (Williamson 1985, 1996) provides a relevant theoretical framework to study the variety of coordination modes and the determinants of the choice between these various modes. It identifies three main modes of coordination: markets, vertically integrated firms and hybrid arrangements, which differ in their ability to provide incentives and to adapt to hazards as well as in the coordination mechanisms they rely on. We focus here on modes of coordination between independent partners (i.e. market and hybrid forms) and leave aside integrated firms, which are beyond the scope of our analysis.

For transactions involving standardized assets, alternative purchase and supply arrangements are presumably easy to find so that market is the efficient mode to coordinate exchanges. In such circumstances where assets are generic, the threat of changing partner is indeed credible as partners are interchangeable, which provides a strong incentive not to behave opportunistically. Price is then a sufficient mechanism to coordinate exchange. In case of exogenous disturbances or litigation, partners can rely directly on the courts and legal rules.

However, problems with markets arise as bilateral dependencies build up, that is to say when the identity of the exchange parties matters so that they wish to ensure the continuity of their relationship, because, for instance, they have invested in relationship-specific assets.[2] To protect the value of these investments and avoid the risks of opportunism that are associated with them, while preserving their autonomy, exchange partners may opt for relational contracts, that is hybrid forms. These modes of coordination, which can take various forms (e.g. subcontracting, franchises, licenses, alliances, etc.), have common characteristics (Ménard 2004). Firstly, they involve the pooling of resources by partners, who are simultaneously legally independent and engaged in market relationships. Secondly, hybrid forms

---

[2] Like, for instance, investments in human capital to ensure the transfer of know-how to the other partner.

rely on contractual arrangements which maintain competitive pressure among participants. Finally, the fundamental characteristic of hybrid forms is their reliance on private mechanisms of coordination and conflict resolution (e.g. collective institutions, intermediation mechanisms, codes of ethics, self-regulation implemented by professional federations, patent brokers).

Transaction cost economics suggests that, for a given institutional environment, the trade-off between these alternative modes of organization depends on the characteristics of the assets involved in carrying out transactions, as they determine the level of contractual completeness. Thus, transactions involving highly specific assets and characterized by a high level of uncertainty about the conditions that prevail during the execution of the contract (whether these conditions are exogenous, i.e. concern the future states of nature, or endogenous, i.e. come from agents' strategic behaviours) are sources of contractual difficulties and are therefore unlikely to be realized on markets. Markets are indeed considered as efficient for transactions involving generic assets for which there is a large number of suppliers and buyers and a low degree of uncertainty. Conversely, in situations of small numbers of exchangers, characterized by bilateral dependency between the suppliers and the sellers and a high level of uncertainty, the costs of using the market (i.e. the transaction costs) are potentially very high due to risk of opportunism (hold up) and maladaptation of the contractual arrangements supporting the transactions. If the autonomy of the parties involved in such transactions can be maintained, the use of hybrid forms, ensuring the adaptation of the relationship through private governance mechanisms, will be preferred.

This typology provides a first means of interpreting the patterns of knowledge transfer, which are characterized by a great diversity. The acquisition of knowledge can of course be done internally by the integration of R&D activities. It can also be done via market transactions such as the purchase of licenses and patents. Finally, knowledge transfer may be the result of partnership agreements, alliances or cross-licensing between competing firms, which correspond to hybrid forms of coordination.

As shown by Lamoreaux and Sokoloff (2001), these various modes of coordination have existed for decades. But, although internal R&D used to be the dominant model of innovation acquisition, since the 1990s, market and hybrid transactions for technologies have grown sizeably due to the increasing complexity of many technical products and systems (Tietze 2012).

3.2.2 Auctions and intermediation

The choice between market and hybrid forms echoes the trade-off between different mechanisms of selection of co-contractors studied by the literature on 'market design'. This approach focuses on the effects of asymmetries or gaps in information relating to the traded goods or to the agents themselves. Indeed the characteristics of the traded goods may or may not be known by buyers, sellers may ignore buyers' willingness to pay, their reputation may be uncertain, and these informational imperfections are at the heart of the optimal choices of market mechanisms.

Without being exhaustive, we come back to the main mechanisms of market transactions that are well suited to situations where uncertainty is an essential component of the traded good as well as of the agents involved in the exchange process. Auctions belong to this category of mechanisms, as they allow the disclosure of information; intermediated markets, where specialized agents agree to realize investments to get information about the goods and the market participants, also belong to this category.

Auction mechanisms

Auction mechanisms are traditionally used to organize transactions on items of which the value is uncertain. Two sources of value uncertainty can be distinguished.

Firstly, the value of the item may depend on the willingness to pay of each potential buyer and may be, a priori, unknown to the seller. This is the case for goods with a marked character of uniqueness (art, real estate, radio spectrum). Unaware of the buyers' willingness to pay, the seller may use different auction mechanisms that allow him both to identify the value of the good and to attribute it to the buyer who is willing to pay the highest price. This implicitly describes a situation of private-value auctions: the value of the good may be unique to each bidder, as it depends on their private use of it. In a way, what creates uncertainty about the value of the good is that the seller does not know the values assigned to the auctioned good by the potential buyers. The distribution of the private values among the population of buyers and the auction rule chosen by the seller jointly determine the price of the auctioned good. One of the remarkable features of this mode of marketing is that it reveals the willingness to pay of buyers, who are thus competing for obtaining the good.

Secondly, the intrinsic value of the good may also be ignored by the buyers, who will discover it only afterwards, if they win the auction. In such auctions, called common-value auctions, all bidders have roughly the same valuation (e.g. the market value per barrel of oil extracted under an oil lease) but they may have different, privately held information (e.g. individuals may hold different interpretations of available data indicating the quantity of oil that may be available in a given lease area). It is in this type of auction that the 'winner's curse' phenomenon is likely to occur: the winner of the auction is the most optimistic bidder; since the auctioned item is worth the same to all bidders, who are distinguished only by their respective estimates, then the winner is the bidder who overestimates the item's value and he is likely to overpay it.

Intermediation mechanisms

For transactions characterized by a high level of uncertainty, an alternative way to connect buyers and sellers is intermediation, which is particularly suited to search goods, that is to say, goods for which buyers are willing to incur information costs (e.g. real estate, insurance, art, financial products).

The role of intermediaries is twofold. On the one hand, their function is to appraise the quality of goods. Insofar as they attend a large number of transactions, in contrast to buyers and sellers who usually buy or sell one good, intermediaries have incentives to invest in the expertise and the knowledge required to estimate the value of goods (see Biglaiser 1993; Biglaiser and Friedman 1994; Lizzeri 1999). This is the case of estate agents for example, or art experts.

On the other hand, the role of expert intermediaries is linked to the two-sided characteristic of the markets on which they operate: it is to achieve the best matching between the buyer, the seller and the adequate item, through their better knowledge of transaction opportunities.

In other words, the dual function of intermediaries consists in identifying the characteristics of the goods and the best potential partners in order to enable the achievement of mutually beneficial exchanges.

The size of the markets encompassed by the intermediaries is therefore likely to be crucial to their efficiency. Indeed, on the one hand, the volume of intermediated goods provides the expert with a better knowledge of the market and, a stronger reputation, and finally it allows a more accurate estimate of the value of each good. On the other hand, being in contact with many participants, on one side (purchases) or the other (sales) of the market, obviously allows achieving a better matching for each transaction. The size of intermediated marketplaces is thus a key determinant of the efficiency of such platforms, as it impacts on their ability to generate network externalities (see Evans 2003; Rochet and Tirole 2003; Armstrong 2004). At the same time however, the size of intermediated marketplaces may not grow indefinitely; it is bounded by the cognitive capacities of intermediaries, especially when the intermediated transactions rely on complex assets, like patents, which require specialized knowledge (Lamoreaux and Sokoloff 1999, 2001; Monk 2009; Tietze 2012).

3.2.3 The choice of optimal coordination

A small number of articles, all very recent, have investigated the endogenous choice among different exchange forms like centralized markets, of which auctions are the prototype, intermediated places, bilateral negotiations over the counter (OTC), etc. This choice can be analysed either from the seller's perspective or from the perspective of all the agents participating in the market.

These studies identify two sets of determinants of the trade-off between the various modes of coordination: the characteristics of the agents on the one hand, and the characteristics of the traded goods on the other hand. We briefly review some of these contributions by focusing on those that seem better to approach the issues raised by the patents markets (see Kugler *et al.* (2006) for a review of the scarce theoretical and experimental literature on the endogenous formation of markets).

Regarding the incidence of traders' characteristics, Kugler, Neeman and Vulkan (2006), as part of an experimental method, and Neeman and Vulkan (2010), as part of a theoretical approach, consider the choice between centralized market and bilateral negotiations in a setting where goods are homogeneous and privately informed buyers and sellers differ, respectively, in their willingness to pay and in their production costs. Each agent can choose between participating either in an auction market or in decentralized negotiations. The results of both experimental and theoretical studies show that, under fairly general assumptions, buyers with a high willingness to pay and sellers with low costs are attracted by centralized markets. Indeed, the extent to which their high willingness to pay and low cost are translated into higher and lower prices, respectively, is smaller in centralized markets, where price-taker agents face a unique market clearing price. As a consequence, the equilibrium of decentralized transactions is unstable, because the switch of a small number of 'serious' buyers or sellers to a centralized mode of coordination pulls the remaining serious traders and ultimately leads to the unraveling of trade through direct negotiations, as there is no more possibility of mutually beneficial exchanges through OTC transactions for the traders, whatever their type.

Rogo (2009) obtains opposite results in an empirical paper where he examines the choice between auctions and bilateral bargaining procedures, in the particular case of mergers and acquisitions, where buyers are uncertain about the target firm's value and sellers incur significant costs to disclose relevant information to buyers as information may be proprietary. Hence, unlike the situation studied by Kugler *et al.* (2006) and Neeman and Vulkan (2010), where it is the characteristics of the exchange partners which are the subject of private information, in Rogo (2009), it is the characteristics of the target companies, i.e. of the 'traded good', that are observable only by the sellers. The results of his econometric tests based on a sample of 400 takeover transactions indicate that the choice between auctions and negotiations depends on the level of risks induced by the disclosure of the information required by buyers to evaluate uncertainty. In other words, sellers compare the benefits of inducing higher competition among bidders in an auction with the costs associated with the disclosure of private information. Indeed, the disclosure of private information is necessary to reduce uncertainty about the good's value and thus increase bidders' participation, but, at the same time, information disclosure can be the source of proprietary leakages as in auctions the information disclosed is inevitably available to third parties, and in particular to targets' competitors. Hence, according to Rogo (2009), the optimality of auctions, which is the result obtained by Kugler *et al.* (2006) and Neeman and Vulkan (2010), ceases to hold when the valuation uncertainty associated with some proprietary information is too strong. Bilateral negotiations may then be the preferred mode of coordination.

The results obtained by Gau and Quan (1992) also emphasize the importance of informational configuration in the choice of a coordination mechanism. They propose a model of mechanism choice in the disposition of real estate assets, considering two alternatives: search or negotiated sales (i.e. bilateral negotiations) and auctions (i.e. centralized markets). Traded goods (real properties) are described as search goods: buyers face search costs while sellers incur costs when they fail to sell their property. The central result of their model is that buyers with high search costs prefer auctions because the auction payoff imposes an upper bound on buyers' gain from search. Instead, buyers with low search costs prefer to turn to bilateral negotiated exchanges. The authors complement this analysis with an empirical test for price differences due to mechanism choice and show that negotiated transactions prices yield higher prices.

The results of these works emphasize the influence of the characteristics of the traders (their willingness to pay, their production costs), and of the costs induced to mitigate uncertainty on the value of the traded goods (costs associated with the disclosure of proprietary information on the side of the seller, search costs on the side of the buyer). However, they do not investigate the characteristics of the goods that are more suitable for being auctioned or negotiated.

The impact of the characteristics of the traded goods on the efficiency of alternative coordination mechanisms is the central issue analysed by Bajari *et al.* (2006, 2009) in their empirical studies on procurement contracts. To investigate whether the buyer of a customized good should use competitive bidding or bilateral negotiation to select a contractor, they draw on insights from the existing theoretical literature on incomplete contracts and transaction costs and test their propositions on a data set of private sector building contracts. They show that, for complex traded goods, whose specifications are hard to complete *ex ante* and for which *ex post* adaptations are expected, that is to say when complete contracting is prohibitively costly, the use of market coordination mechanisms (competitive tendering in their case) leads to costly renegotiations and harmful strategic behaviour by bidders. Indeed, a risk incurred when auctions are used for complex items is the increase of the bidding costs. If the buyer fails to specify the subject matter of the bid with precision, then uncertainties will result, costs of bidding will be increased, and applicants will be discouraged. The number of bidders being limited, the expected benefits of competitive tendering would consequently be affected. Or the number of bidders may not be limited but, since they anticipate future renegotiation due to contractual incompleteness, their bid may incorporate high risk premiums for them to be able to recover potential adaptation costs.[3]

In addition, if the description of the project or the auctioned good is not sufficiently clear, competitive tendering may lead to situations of adverse selection and end up with the selection of the most opportunistic bidder. If contractual design is incomplete and the good or the service is complex, an auction may indeed lead to choosing the bidder who is the most aware of the contractual blanks he could exploit, that is to say the one who is able to determine where contracts will fail. Anticipating that he will be able to take advantage of situations that are unforeseen in the contract by renegotiating the initial arrangement, this strategic candidate will not hesitate to propose an unrealistically low price. This type of bidding behaviour (low-balling strategy) jeopardizes allocative efficiency, which is the most important objective of tendering. In such circumstances, the use of negotiated procedures and relational contracts is recommended because they allow the auctioneer (in this case the buyer) to acquire information on the characteristics of the object and adapt the contractual arrangement to changing circumstances.

In short, it is clear from these analyses that in the presence of uncertainty, when information is incomplete or asymmetrically distributed between the participants and when the good or service exchanged encompasses several dimensions that are hardly contractible, price can no longer be an effective coordination mechanism since it does not reflect the relevant data and information on the market. Complementary mechanisms of coordination and information acquisition are then required.

3.3 Characteristics of patents

The analysis of the characteristics of patents and of the informational configuration of the markets for knowledge is a necessary step to be able to make propositions on the ability of markets to coordinate patents transactions.

---

[3] In their study of highway construction and maintenance contracts in California, Bajari *et al.* (2006) estimate these risk premiums to represent, on average, 10% of the value of the contract.

Hence we present, firstly, the characteristics of patents which make it difficult to estimate their value. This uncertainty of the value of patents is widely stressed in the literature but also by R&D practitioners, who struggle to converge to a clear and reliable estimation method (see Baudry, Chapter 5). In a second step, we recall that uncertainty also comes from the environment of the transactions, that is to say from the characteristics of buyers and sellers but also from the quality of institutions that guarantee the respect for intellectual property rights.

3.3.1 Uncertainty on the intrinsic value of patents

The 'intrinsic' value of a patent is subject to uncertainty. Several reasons may explain the difficulty that both buyers and sellers may experience to evaluate patents and therefore to use the market to exchange them.

Patents may incorporate a more or less large proportion of unprecedented innovation. They can partially cover previous innovations so that their marginal contribution is limited. Further, many patents issued are questionable or 'weak' and might well be found invalid if litigated.[4] Hence, as stressed by Lemley and Shapiro (2005), patents are probabilistic property rights and the innovation they convey is often known *ex post*, so that patents fall into the category of experience goods. Indeed, it is often difficult to assess *ex ante* the market value of an invention that is likely to lead to the creation of entirely new products for which demand is by definition unknown. In auction models, such a situation characterizes common-value auctions, in which the value of the auctioned good is identical for all bidders but unknown at the time of the auction/sale. In this situation, one limitation of the use of auctions, that is to say of a coordination by prices, is that buyers, anticipating a risk of 'winner's curse', seek to mitigate it by minimizing their offers (Hong and Shum 2002).

Another consequence of the uncertainty on the intrinsic value of patents is that buyers may anticipate a risk of adverse selection of the patents that are put on the market. Indeed, insofar as a patent holder

---

[4] This is due to the fact that it would be very costly for a patents and trademarks office (PTO) to scrutinize all patent applications as thoroughly as courts examine the relatively few litigated patents (Lemley 2001).

may either keep the patent to exploit it or may sell it, it is rather likely that only the less valuable patents (the worst, the less exploitable, those protecting the least interesting innovation . . .) are for sale. Anticipating this mechanism of adverse selection, buyers have little incentive to enter the market and, if they do so and decide to participate, they are likely to offer very low prices. As shown by Akerlof (1970) in his article on the market for 'lemons', this type of effect can completely make the market disappear since, ultimately, sellers, anticipating the withdrawal of buyers or their very low valuation of the patents offered for sale, will not put their patents up for sale. However, certain patents offered for sale are of good quality, but they suffer from the fact that the antici-pated average quality is low.

In addition, a patent may be exploitable, and therefore valuable, only if it comes with many other complementary patents, as in the case of technological standards. In sectors that are particularly technology intensive, such as electronics and biotechnologies, it is rare that a single patent enables the production of a good in the downstream market, which may induce patent thicket phenomena, as described by Encaoua and Madiès (Chapter 6). Therefore, the value of a particular patent can vary considerably depending on whether the complementary patents are available or not; its price will thus depend on whether the pur-chaser already holds the complementary patents or can acquire them on the market. Otherwise, if the complementary patents require costly research with an uncertain outcome, then the value of the initial patent is reduced accordingly. Thus, the value of a patent depends on the availability of complementary patents, on the identity of the buyer and on its endowment of complementary patents.

Similarly, in some cases, a patent may be valuable only if the transfer of technology it induces is accompanied by a transfer of the know-how of the inventor. The ability of buyers to appropriate the gains from innovation is indeed closely linked to the know-how embedded in the technology. But because know-how is sometimes tacit, its evaluation and its contractibility may be difficult, which leads to the development of relational arrangements (partnerships, joint venture).

3.3.2 Uncertainty on agents and transactional environment

Regardless of the actual content of a patent, the environment in which it is likely to be exchanged may contain many factors of uncertainty.

First, the characteristics of the exchange partners are likely to introduce additional uncertainty on the value of the transactions. Thus, a seller may not be indifferent to the identity of the agent who buys his patent. The latter's position in the market may allow him to benefit more or less from the applications enabled by the patent. The degree of competition and rivalry between the competitors of the buyer on the downstream markets, where those goods incorporating innovation are sold, appears to be essential in determining the buyer's willingness to pay. Whoever has the highest willingness to pay for the patent is the one who will value it the best in the downstream market by incorporating the innovation in goods or services that are particularly profitable. But this buyer, because he is perhaps the most efficient in the downstream market, or the one whose brand has the best reputation, is probably the most threatening competitor to the seller. The seller thus faces a trade-off between the revenue from licensing on the upstream market with the rent-dissipation effect created by the fact that the license creates a competitor in the firm downstream market, or it makes the competitor more efficient.[5] In an auction market, this feature reflects a private-value auction: the value placed on a particular good depends on the identity of the potential buyer.

On the other hand, there may be uncertainty about the quality of patent redaction. The latter can indeed be written in a more or less accurate way and result in litigation, which can be interpreted as a reduction in quality for the buyer. Now, given that the likelihood of lawsuits is obviously related to commercial stakes and to the maturity of the technological domain (Lanjouw and Schankerman 2001), it also depends on the quality of the reviewing process of patent applications by patents offices (Zorina Khan and Sokoloff 2001; Caillaud and Duchêne 2011) but also on the mode of remuneration of patent lawyers (Duchêne 2008).

---

[5] This trade-off between revenues generated by the sale of patents (or patents licensing) and rent dissipation has been studied by Arora and Fosfuri (2003) who show that licensing is discouraged if: (i) the downstream operations of the firm are large because a firm with larger sales has more to lose from the entry or the higher efficiency of a competitor (the rent dissipation effect is then stronger), (ii) the downstream market is more differentiated (for the same reason as before) and (iii) the intellectual property rights are weak, which reduces the expected revenues from licensing.

More broadly, the quality of the institutional environment in which patent transactions are made is a key determinant in the choice of a mode of coordination. Specifically, institutions guaranteeing the respect of intellectual property rights (IPR) and contractual commitments facilitate the development of market transactions as they reduce the costs of exchange (Teece 1986; Anton and Yao 1994). On the contrary, if the institutional environment does not insure against the risks of expropriation and does not ensure the credibility of contractual commitments, transactions must be governed by private ordering arrangements created by the agents themselves (Bessy, Brousseau and Saussier 2002).

The importance of institutional quality is illustrated by Lamoreaux and Sokoloff (2001) and Zorina Khan (1995) in their series of works on the history of innovation and patent transactions in the United States. The authors show that the switch in 1836 from a patent registration system, which left to civil litigation the resolution of questions of originality, novelty and appropriate scope, to an examination system in which technically trained examiners scrutinized each application to ensure that the invention constituted an original advance in the state of the art, was followed by an exponential growth of patents transactions.

Similarly, the studies by Oxley (1997, 1999) on technology transfer agreements made by American firms in foreign countries have shown that the type of agreement chosen by firms depends on the type of technology that is transferred but also on the IPR protection regime which exists in the partner country, a result corroborated by the study by Hagedoorn, Cloodt and van Kranenburg (2005) on a larger set of international inter-firm R&D partnerships. More precisely, these studies, which focus on the trade-off between equity joint venture and contractual partnership, show that companies prefer equity-based partnerships to contractual agreements when they are confronted with higher levels of specific knowledge transfer. Above all, it is shown that in countries where intellectual property is poorly protected, firms prefer hybrid modes of coordination (e.g. joint ventures) rather than contractual agreements because the former offer the safeguards against appropriability hazards (e.g. technology leakage) that institutions do not provide. In other words, when uncertainty surrounding partnerships increases, due to the high specificity of the assets and technologies at stake or due to the poor quality of institutions, market mechanisms are not sufficient to coordinate exchanges.

To summarize, patents simultaneously possess all the characteristics that make it difficult to use the market and the price mechanism to coordinate exchanges. These are common- and private-value items: their intrinsic value is uncertain *ex ante*, and it also depends on the identity of the buyer, specifically on the opportunities he has on the downstream markets. Patents are search goods and experience goods: estimating their value requires incurring costs of research that may not be enough because the potential of the innovation protected by the patent can only be found *ex post*. Finally, their value may be contingent on the existence of complementary assets (other patents or know-how). Also, the high degree of uncertainty surrounding patent transactions leads us to question the potential for development of markets for patents.

3.4 What are the lessons for patent markets ?

3.4.1 Limits of price signal and auctions

The use of auctions seems to be conceivable in specific technology areas, where knowledge can easily be codified. The codification of knowledge indeed plays a major role in the development of market transactions to the extent that it allows one to identify more clearly the object protected by the patent, thus reducing transaction costs.

Overall, the development of patent markets in which transactions would be made on the basis of the sole price signal seems to be limited. As emphasized by Anand and Khanna (2000), Bessy, Brousseau and Saussier (2002) and Jarosz *et al.* (2010), the multiplicity and complexity of the contractual dimensions to be considered to transfer patents and licenses lead agents to favour relational coordination. These authors show that technology licensing agreements are far from standard and do not constitute a homogeneous set of arrangements. These are rarely 'pure' market contracts based solely on price information but rather relational contracts providing for various remuneration schemes, mechanisms for dispute resolution but also various mechanisms to facilitate knowledge transfer.

Given the limited scope of auction mechanisms in the field of intellectual property, and the agents' preference for relational arrangements, it seems more appropriate to use intermediation mechanisms to facilitate the development of patents transactions.

3.4.2 Securing transaction with intermediation mechanisms

As already pointed out, the execution of transactions and the gains obtained from patent exchange depend on the informational pattern prevailing in the market. What is crucial in the informational configuration is, firstly, the nature of available information on the intrinsic characteristics of the patent, and secondly, the nature of the information that players hold about themselves and about the other.

In this context, experts have a central informational and 'reassuring' function. For items such as patents, with an uncertain intrinsic value and prices that vary depending on the identity of the buyer, intermediaries may establish links between the seller and the buyer who is the most likely to buy the patent while allowing them to make custommade relational arrangements (Howells 2006; Di Minin and Benassi 2008; Hoppe and Ozdenoren 2008).

One advantage of intermediation is to reduce the costs of finding a partner, which can be prohibitive in the domain of intellectual property, both for sellers and buyers. Indeed, it may be difficult for sellers to find appropriate partners without incurring the cost of advertising the availability of their patents. For their part, potential buyers do not often want their desire to acquire a patent to be public for fear of revealing the weakness of their legal or competitive patent portfolio.

In addition, intermediaries reduce the costs associated with bilateral negotiations that both sides of the market support as well as the opportunity costs associated with devoting a substantial portion of a patent's limited life to patent negotiation rather than patent exploitation.

Moreover, the fact that experts have to build their own reputation can negate the issue of adverse selection that was mentioned above. Indeed, while the seller involved in a single transaction can hardly build a solid reputation, an expert who certifies the quality of a patent is credible if its assessments in the past were relevant. The presence of an intermediary acting as an experienced assessor prevents the possible disappearance of the market which can occur when guarantees are indefinite (Akerlof 1970).

One condition for this to hold is that intermediaries have the means to collect information more efficiently than exchange participants themselves and that their presence allows the realization of beneficial exchanges. Specialization of intermediaries in a given field of science or technology can help, as illustrated in the article by Monk (2009) on the recent emergence of patent intermediaries in the Silicon Valley. In the same vein, Lamoreaux and Sokoloff (1999) show a strong correlation between the geographical location of innovation and the location of intermediation places. On the one hand, institutions of intermediation in patented technologies are concentrated in areas where the rate of invention is already high. On the other hand, the presence of firms and institutions facilitating the exchange of inventions in an area stimulates a greater specialization and greater productivity of inventions in this area because it increases the net returns that inventors can expect and allows them to raise capital to support their innovation activity more easily.

3.5 Conclusion

The lessons from the literature on the optimal choice of patents exchange procedures do not suggest the intrinsic superiority of one particular coordination mode in all circumstances. Indeed, most of the studies we have presented highlight the very high sensitivity of the optimal mechanism of patents transfer to the informational configuration prevailing on the market, that is to say, to the nature and distribution of the information available on patents and agents. Moreover, the most efficient coordination mode also depends on the technological domain and on the scientific environment in which innovations are made because they affect the capacity to codify knowledge. It is also dependent on the competitive environment in which agents evolve, and in particular on the existence of downstream competition between buyers and sellers of patents.

Insofar as there is not one unique efficient mode of coordination of patents exchange but several modes, there cannot be one unambiguous action for policy-makers. Encouraging the development of expertise on patents can certainly be an efficient action to promote exchanges, but, if rents exist in such an activity, then patents intermediaries should appear spontaneously, as seems to be the case in some industries and some geographic areas (e.g. in the field of electronics in the Silicon Valley).

References

Akerlof G. A. (1970) The market for 'lemons': quality uncertainty and the market mechanism. *Quarterly Journal of Economics*, **84**, 488–500.

Anand B. N. and T. Khanna (2000) The structure of licensing contracts. *Journal of Industrial Economics*, **48**, 103–135.

Anton J. and D. Yao (1994) Expropriation and inventions: appropriable rents in the absence of property rights. *American Economic Review*, **84**, 190–209.

Armstrong M. (2004) Competition in two-sided markets. *RAND Journal of Economics*, **37**, 668–691.

Arora A. and A. Gambardella (2010) Ideas for rent: an overview of markets for technology. *Industrial and Corporate Change*, **19**, 775–803.

Arora A. and A. Fosfuri (2003) Licensing the market for technology. *Journal of Economic Behavior and Organization*, **52**, 272–295.

Athreye S. and J. Cantwell (2007) Creating competition? Globalisation and the emergence of new technology producers. *Research Policy*, **36**, 209–226.

Bajari P., S. Houghton and S. Tadelis (2006) *Bidding for Incomplete Contracts: An Empirical Analysis*, NBER Working Paper 1201. Cambridge, MA: National Bureau of Economic Research.

Bajari P., R. McMillan and S. Tadelis (2009) Auctions *versus* negotiations in procurement: an empirical analysis. *Journal of Law, Economics and Organization*, **25**, 372–399.

Bessy C., E. Brousseau and S. Saussier (2002) *The Diversity of Technology Licensing Agreements*, mimeo, University Paris X.

Biglaiser G. (1993) Middlemen as experts. *RAND Journal of Economics*, **24**, 212–223.

Biglaiser G. and J. W. Friedman (1994) Middlemen as guarantors of quality. *International Journal of Industrial Organization*, **12**, 509–531.

Caillaud B. and A. Duchêne (2011) Patent office in innovation policy: Nobody's perfect. *International Journal of Industrial Organization*, **29**, 242–252.

Di Minin A. and M. Benassi (2008) *Playing in Between: Patents' Brokers in Markets for Technology*, Working Paper 200802 Pisa: Sant'Anna School of Advanced Studies.

Duchêne A. (2008) *How Lawyers Protect Innovation: The Rules of the Double-game*, working paper. Philadelphia, PA: Drexel University.

Evans D. (2003) The antitrust economics of multi-sided platform markets. *Yale Journal of Regulation*, **20**, 325–382.

Gambardella A., P. Giuri and M. Mariani (2005) *The Value of European Patents: Evidence from a Survey of European Inventors*, Final Report of the PatVal-EU Project. Pisa: Laboratory of Economics and Management (LEM), Sant' Anna School of Advanced Studies. www.alfonsogambardella.it/patvalfinalreport.pdf

Gans J. and S. Stern (2010) Is there a market for ideas ? *Industrial and Corporate Change*, **19**, 805–837.

Gau G. and D. Quan (1992) *Market Mechanism Choice and Real Estate Disposition: Negotiated Sales versus Auctions*, mimeo, University of California Los Angeles, Anderson Graduate School of Management. http://escholarship.org/uc/item/77f5k3x9.

Hagedoorn J., D. Cloodt and H. van Kranenburg (2005) Intellectual property rights and the governance of international R&D partnerships. *Journal of International Business Studies*, 36, 175–186.

Hendel I., A. Nevo and F. Ortalo-Magné (2009) The relative performance of real estate marketing platforms: MLS versus FSBOMadison.com. *American Economic Review*, 99, 1878–1898.

Hong H. and M. Shum (2002) Increasing competition and the winner's curse: evidence from procurement. *Review of Economic Studies*, 69, 871–898.

Hoppe H. C. and E. Ozdenoren (2008) Intermediation in innovation. *International Journal of Industrial Organization*, 23, 483–503.

Howells J. (2006) Intermediation and the role of intermediaries in innovation. *Research Policy* 35, 715–728.

Jarosz J., R. Heider, C. Bazelon, C. Bieri and P. Hess (2010) Patent auctions: how far have we come? *Les Nouvelles*, March 2010, 11–30.

Kugler T., Z. Neeman and N. Vulkan (2006) Markets versus negotiations: an experimental investigation. *Games and Economic Behavior*, 56, 121–134.

Lamoreaux N. R. and K. L. Sokoloff (1999) The geography of the market for technology in the late-nineteenth- and early-twentieth-century United States, in Gary D. Libecap (ed.) *Advances in the Study of Entrepreneurship, Innovation, and Economic Growth*, vol. 11, pp. 67–121. Bingley, UK: Emerald Group Publishing.

(2001) Market trade in patents and the rise of a class of specialized inventors in the nineteenth-century United States. *American Economic Review*, 91, 39–44.

Lemley M. A. (2001) Rational ignorance at the patent office. *Northwestern University Law Review*, 95, no. 4.

Lanjouw J. O. and M. Schankerman (2001) Characteristics of patent litigation: a window on competition. *RAND Journal of Economics*, 32, 129–151.

Lemley M. A. and C. Shapiro (2005) Probabilistic patents. *Journal of Economic Perspectives*, 19, 75–98.

Lizzeri A. (1999) Information revelation and certification intermediaries. *RAND Journal of Economics*, 30, 214–231.

Ménard C. (2004) The economics of hybrid organizations. *Journal of Institutional and Theoretical Economics*, 160, 345–376.

Mendi P. (2007) Trade in disembodied technology and total factor productivity in OECD countries. *Research Policy*, 36, 121–133.

Monk A. H. B. (2009) The emerging market for intellectual property: drivers, restrainers and implications. *Journal of Economic Geography*, 9, 469–491.

Neeman Z. and N. Vulkan (2010) Markets versus negotiations: the predominance of centralized markets. *The B.E. Journal in Theoretical Economics*, 10, 6–16.

Oxley J. E. (1997) Appropriability hazards and governance in strategic alliances: a transaction cost approach. *Journal of Law, Economics and Organization*, 13, 387–409

(1999) Institutional environment and the mechanisms of governance: the impact of intellectual property protection on the structure of inter-firm alliances. *Journal of Economic Behavior and Organization*, 38, 283–309.

Rochet J.-C. and J. Tirole (2003) Platform competition in two-sided markets. *Journal of the European Economic Association*, 1, 990–1029.

Rogo R. (2009) *The Effect of Valuation Uncertainty in the Choice of Selling Mechanism*, mimeo, Kellogg School of Management, Northwestern University, Evanston, IL.

Tapscott J. D. and A. D. Williams (2008) *Wikinomics: How Mass Collaboration Changes Everything*. London: Penguin.

Teece D. J. (1986) Profiting from technological innovation: Implications for integration, collaboration, licensing and public policy. *Research Policy*, 15, 285–305.

Tietze F. (2012) *Technology Market Transactions: Auctions, Intermediaries and Innovation*. Cheltenham, UK: Edward Elgar.

Troy I. and R. Werle (2008) *Uncertainty and the Market for Patents*, working paper 08/2. Cologne, Germany: Max Planck Institute for the Study of Societies.

Williamson O. E. (1985) *The Economic Institutions of Capitalism: Firms, Markets, Relational Contracting*. New York: Free Press.

(1996) *The Mechanisms of Governance*. New York: Oxford University Press.

Zorina Khan B. (1995) Property rights and patent litigation in early nineteenth-century America. *Journal of Economic History*, 55, 58–97.

Zorina Khan B. and K. L. Sokoloff (2001) The early development of intellectual property institutions in the United States. *Journal of Economic Perspectives*, 15, 233–246.

4 | Structuring the market for intellectual property rights: lessons from financial markets

OLIVER GASSMANN, MARTIN A. BADER
AND FLORIAN LIEGLER

4.1 Introduction

Intellectual property rights (IPR) have become a valuable economic commodity in the knowledge economy, gaining in importance as a strategic competitive advantage. Furthermore, access to IPR is crucial for companies that wish to develop or expand their product range. These factors raise the question of the optimal allocation of IPR, thus leading to the inquiry of the market for IPR. Literature on markets for technology (e.g. Gambardella, Giuri & Luzzi, 2007) and literature on market design (e.g. Roth, 2008) are being merged only recently (Gans & Stern, 2010) and call for further inquiry. This chapter is based on the findings of the study 'Creating a financial market for IPR in Europe' (Bader et al., 2012), building upon previous literature and relying on our own empirical investigation.

Today, companies and research organisations are already trading and licensing patents. However, the existing (unorganised) market lacks transparency, and uncertainty on the quality and value of patents and technology drives up transaction costs. The emergence of an organised IPR market should create value for the participating actors, but should also generate general economic and societal benefits.

Given that the IPR market will develop and the challenges will be overcome, the creation of the IPR market will lead to economic growth driven by increased innovation activity. Economic growth correlates with increased wealth and therefore leads to societal benefits.

Furthermore, a successful organised IPR market will stimulate innovators to turn their ideas into exchangeable and exploitable assets. As a result, it can lead to increased IPR circulation and improved access to IPR. Trading IPR assets on the market will increase the visibility and the competitiveness of the participating actors. It supports the transfer

The authors would like to thank Simon Lapointe for helpful comments.

of ideas from inventors to institutions capable of and experienced in making a new product or process a market success.

The IPR market can be considered a tool to create economic value by allocating knowledge in a wealth-maximising way by offering an incentive to produce knowledge that is broadly accessible to other economic actors in exchange for monetary benefit. Following this theory of markets' efficient allocation, it is also likely that knowledge will be distributed to those actors that can create the largest economic benefit since they have the highest willingness to pay. In many cases, knowledge is tacit, which means that it is bound to people or organisations. Detaching that knowledge in terms of formalisation is costly in many cases. In fact, making knowledge explicit is a question of profitability. The market can stimulate the willingness to make knowledge explicit by offering monetary compensation.

The financial side of the IPR market is likely to develop a range of new financial products aimed at funding innovations. Attracting more capital for innovation is expected to lead to increased research activity. In turn, this will incentivise research institutions and industrial companies to engage in developing innovations. Specifically, this could narrow the funding gap between research and prototype development.

Consequently, the private sector's stronger integration into the production of innovations via the IPR market may increase the efficiency and commercial orientation of research efforts.

In the following section, we will outline the IPR market model to provide the basis for an investigation of the market. Thereafter, we discuss the market's functions and the success factors of financial products that need to be fulfilled. Section 4 contains the main challenges that need to be overcome in the evolution of an organised IPR market, which lays the foundation for successful structuring and trading of financial products in Section 5. The chapter concludes with lessons learned from both policy perspective and company perspective (see Box 4.1).

4.2 Model of the IPR market

An organised IPR market must differentiate between the IPR asset market (where IPR are traded) and the IPR financial market (where investors can invest in IPR). These two markets are the model's basic components. This structure proposes an organised IPR asset market which is connected to the IPR financial market by financial products or vehicles (see Figure 4.1).

Box 4.1 Background and methodology

Chapter background

The findings of this chapter are based on the study 'Creating a financial market for IPR in Europe', which was undertaken on behalf of the European Commission by the Institute of Technology Management at the University of St Gallen in Switzerland and the Fraunhofer Centre for Middle and Eastern Europe in Germany. We worked closely together with industrial firms, IPR holders and European banks. In addition, we conducted a large-scale survey with the most IPR intensive firms in Europe.

The study was conducted in 2011 and has recently been published by the Commission (Bader *et al.*, 2012). The study's final report can be downloaded from the website of the European Commission's Directorate General of Enterprise and Industry at: http://ec.europa.eu/enterprise/policies/innovation/policy/intellectual-property/index_en.htm

The study 'Creating a financial market for IPR in Europe' has been carried out by the contractors on behalf of the European Commission. The study should not be constructed as reflecting the position of the European Commission and its services. Neither the European Commission nor any person acting on behalf of the European Commission is responsible for the use which might be made of information contained therein.

Methodology

In order to grasp the concept of the IPR market, three main research steps have been applied to cover the study objectives.

(a) Literature research

The desktop research was aimed at providing a substantial overview of the concepts involved in establishing an IPR financial market. The results of the literature research form the basis for the definition of the concepts involved in the study and for the conducted expert interviews.

(b) Expert interviews

Interview guidelines were developed on the basis of the desktop research results. These interview guidelines were used to ascertain the interviewees' general view of an IPR asset market and an IPR financial market, as well as the concepts covered in the course of the study. Over 80 face-to-face or telephonic semi-structured interviews were conducted with experts in the field of IPR, technology transfer, markets for patents and technology, patent infringements, patent aggregation and banking.

Box 4.1 *(cont.)*

(c) Empirical study

The empirical study was conducted in three main phases:

(1) The concept development and survey design; developing and structuring the questionnaire; and identifying contact persons.
(2) An online survey of the top 1,000 patent applicants at the European Patent Office (EPO).
(3) Analysing the online survey responses to obtain a better understanding of patent applicants' current and envisioned situation regarding the IPR market.

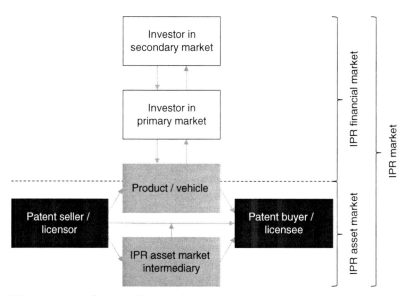

Figure 4.1 Our framework: an indicative market structure.

4.2.1 IPR asset market

When patent vendors or licensors transfer their IPR to patent acquirers or licensees for financial consideration, they do so either directly or indirectly. In the latter case, vendors and acquirers utilise the specialist know-how of intermediaries. The value of these intermediaries lies in identifying transactional partners and providing services such as portfolio management and valuation know-how. On the IPR asset market, therefore, transactions concern patents and licenses directly, as opposed to the IPR financial market.

Even though early stages of IPR asset market design have been tested (e.g. Ocean Tomo organised patent auctions, which have been held in San Francisco since 2006), there is still no generally accepted model operating yet. The IPR asset market is still characterised by a relatively low level of homogeneity due to certain characteristics of patents, such as their context dependence and the additional tacit knowledge required for their successful application. Every patent is unique and non-interchangeable. Moreover, the IPR asset market lacks a generally accepted valuation method and is highly fragmented.

The concept of an organised IPR market aims to address these inefficiencies and lay the foundations for the broad, financially driven capitalisation of innovation.

4.2.2 IPR financial market

In the IPR financial market, vehicles are used to create financial products based on IPR. These vehicles connect the IPR asset market and the IPR financial market and are described in Section 2. The primary market is created when such vehicles issue shares, bonds or hybrids. A secondary market is established as soon as these financial products can be traded between investors.

The financial side of the IPR market directly connects capital to innovation, thus promoting research activity and incentivising companies and research institutions to develop inventions. This effect could narrow the funding gap between the research stage and the development of prototypes. Thus, the market provides for liquidity and homogeneity of assets and resolves the issue of uncertain valuation. It provides understandable products, cash-flow predictability and an organised structure (preferably based on an electronic platform), enabling financial products based on IPR to be traded on a large scale.

4.3 Structuring the IPR market

To ensure efficient resource allocation, financial markets fulfil three different economic functions: determining the product price, offering liquidity and reducing transaction costs (Fabozzi, Modigliani & Ferri, 1994). We find that two of these (liquidity and low transaction costs) are

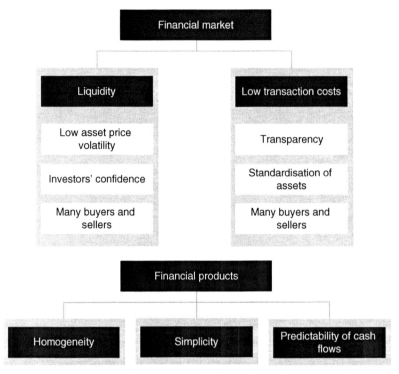

Figure 4.2 Financial market criteria and financial product criteria from our analysis.

key success factors, since the determination of the product price, or the price discovery process, is rather a consequence of the two other functions. In addition, the literature suggests further success factors inherent in financial products (homogeneity, simplicity and predictability of cash flows) that also need to be considered in the evolution of an organised IPR market. Figure 4.2 summarises the success factors of financial markets and financial products as described below.

4.3.1 Liquidity

Although frequently cited as a requirement of a functioning market, the concept of liquidity is not easy to grasp. O'Hara (2003), for example, states that liquidity constitutes a situation in which supply matches demand, i.e. in which buyers and sellers come together with complementary aims. Black (1971), with another viewpoint, defines the term via small bid–ask spreads – the difference between the highest buy and the lowest sell offer. For investors, the focus is on being able to sell an asset whenever they wish to without having to accept significant discounts on the price.

The extensive trade of financial instruments only takes place when these instruments can be exchanged with a certain ease. Amihud and Mendelson (1991) list several costs associated with illiquid markets which inhibit trade. Therefore, liquidity and the existence of secondary markets are closely related. On the one hand, if assets can be traded easily, a market is likely to develop. On the other hand, an existing market structure can facilitate the exchange of instruments. Consequently, liquidity and secondary markets foster as well as require each other. The advantages of secondary markets are, however, only valid if they are liquid.

As with the existence of secondary markets, determinants of liquidity (the number of buyers and sellers, investor confidence and volatility) simultaneously cause and are caused by liquidity. This caveat needs to be kept in mind when liquidity has to be created, as its first determinant illustrates: a large number of buyers and sellers make a market more liquid. Empirically, Benston and Hagerman (1974) find an inverse relationship between the number of shareholders and dealers and bid–ask spreads. Chordia, Roll and Subrahmanyam (2000) confirm the correlation between high trading volume and small bid–ask spreads in respect of individual stocks as well as the market. In general, a large number of buyers and sellers are therefore associated with liquidity. In turn, however, a liquid market also provides an incentive for actors to enter (O'Hara, 1995). This circularity has an important implication: if liquidity is to be increased by inducing actors to enter the market, this can be achieved either by providing incentives other than liquidity or, paradoxically, increasing liquidity through another determinant than the one just presented.

The same principle applies to investor confidence. If investors believe in the functioning of a market, they will be inclined to participate in trading. Consequently, the number of market participants rises, contributing to liquidity. Equally, a liquid market leads to smooth transactions and will thus strengthen the investors' confidence in the market. Again, in order to start the process, either liquidity or investor confidence can serve as a trigger.

As a final determinant, the volatility of asset prices has a negative influence on liquidity. Chordia, Sarkar and Subrahmanyam (2005) empirically confirm this finding, identifying a positive correlation between volatility and bid–ask spreads in both the stock and the bond market. However, they also note evidence of causal circularity, meaning that an illiquid market entails more volatile returns.

4.3.2 Transaction costs

Various publications consider low transaction costs to be a defining characteristic of liquid markets. De Haan *et al.* (2009) describe transaction costs as the effort required in terms of time and money to complete a transaction. This clearly links to the definition of liquidity as the ease with which an asset can be exchanged. Transaction costs can thus be viewed as another determinant of liquidity. Nevertheless, their importance and scope justify their separate discussion.

The transaction cost concept goes back to Coase (1960), who mentions that market participants need to find a counterpart and acquire information about the terms and the traded asset. The parties then have to come to a result that is mutually accepted. Dahlman (1979) classifies the different costs involved and concludes that low transaction costs are essential for the functioning of a market.

Transparency plays an important role in terms of finding a counterpart and the necessary information. If the participants, conditions and assets traded in a transaction are disclosed, actors can form an idea of the prevalent market conditions. They know whom to approach when trying to buy or sell an instrument and gain a rough understanding of the prices paid, and can assess different products and their exchange conditions. Extensive analyses prior to an agreement or even an offer become obsolete. By ensuring transparency in a market, search and information costs can therefore be reduced (Pagano & Röell, 1996).

Bargaining and decision costs describe the effort required to come to an agreement. Again, transparency can help to create a 'standard of best practice' adhered to in the market. This transparency would abolish the need to change the way an agreement is closed with each new counterpart, allowing more efficient, and therefore less costly, bargaining. Williamson (1985) points out that the specificity of traded assets is another important determinant of transaction costs. He argues that transaction costs are higher for assets which can be used for fewer purposes, i.e. assets that are confined to a specific function. Instruments with unique features mean that the market participants must spend considerable effort analysing the subject of a transaction each time they engage in trade, leading to high bargaining costs. In order to counter this effect, it is necessary to achieve a certain degree of standardisation (Williamson, 1985). Through this, the process of bargaining is simplified and the costs involved can be reduced.

The ability to trade almost without bargaining in today's financial markets is usually itself associated with costs of market organisers or market makers who assume an intermediating role. Actors such as dealers and brokers enable markets, which require minimal effort in terms of coming to an agreement, but charge for their services. Fabozzi and Modigliani (1996) identify several such costs, for example, brokerage fees. While actors are not readily influenced by the design of financial instruments, the authors point to the frequency of transactions as another determinant of transaction costs. Williamson (1985) argues that in the case of high trade volumes, the costs of introducing the market-making institutions mentioned above can be spread out over a larger number of trades, reducing each transaction's share. Since many buyers and sellers increase the frequency of trades, they are another determinant of bargaining costs.

4.3.3 Homogeneity

Homogeneity means that financial products of a certain class have the same properties. There are no differences in their principal structure. While desirable, it should be noted that interchangeability is not invariably linked to a certain class of financial instruments. It can, however, be influenced to a degree, for example, through pooling (Saunders & Cornett, 2009). The pooling of IPR assets is what some financial products (e.g. patent funds) offer to create more homogeneity. Therefore, investors do not have to sacrifice substantial time and effort to assess a product. This reduces the overall transaction costs and contributes to the quality of a market.

4.3.4 Simplicity

Shen (2009) emphasises that 'The design of [the financial product] should be chosen carefully to make [it] easy to understand and trade.' The simpler a financial product is, the higher is its appeal to the potential market participants. In such a case, the investors' demand increases and, as a consequence, so does market liquidity. Therefore, it is important that not only the structure of the financial product is simple, but also that its content, i.e. the patents owned, are clearly described and their purpose is communicated to the investors.

4.3.5 Predictability of cash flows

Cash-flow predictability means that the cash flows generated by a financial product should follow a certain pattern, which can be assessed beforehand. It is important to note that cash-flow predictability does not exclude returns contingent on events. It merely demands that the relationship between the cause and the resulting cash flow is clearly defined.

This concept can help to reduce the volatility in an instrument's price. The valuation of financial products is usually based on discounted future cash flows. In this context, arbitrary returns that are difficult to assess make it more likely that market participants will be surprised by market developments. Consequently, their estimates of an instrument's value are abruptly revised, causing high fluctuations in the asset's price. If, on the other hand, returns follow a reasonably predictable pattern, investors' expectations do not experience such sudden shocks. Although cash flows may vary, they do so for at least theoretically predictable reasons. The fundamental estimate of a financial instrument's value is therefore adapted, but generally subject to less volatility. This concept is one of the reasons for companies tending to maintain stable dividend payments even when doing so requires additional outside financing. A predictable cash-flow structure could mitigate volatility, contributing to a more liquid market. The predictability of cash flows is generally higher for IPR which cover technologies that are already in use. Therefore, financial products containing IPR relating to early-stage technologies will face more difficulties in communicating their cash-flow predictability to investors than those who consist merely of mature technologies' IPR.

4.4 The challenges

Several challenges have to be overcome before an organised IPR market can be established and function properly, as the concept of an IPR financial market is an entirely new one. Although there have been some attempts to create financial products based on patents, this has taken place only quite recently and experience is still limited.

4.4.1 Access

The market's governance structure should address the arising challenges ideally on a supranational policy level. A European position is also needed on issues such as access and contributions, taxation and market location. Access to IPR is becoming an increasingly crucial factor in global competition. The United States, China, South Korea and other countries have already acknowledged this trend by creating public–private vehicles, aggregating patents and financing innovation. The implications of such intervention for international trade agreements have yet to be fully considered. Moreover, there is scope for policy discussion on the competitive effects that such vehicles may have on technology companies. The governance of the market may present solutions to these issues, but is yet to be developed and must reflect a well-balanced perspective.

Access must also be considered from an internal standpoint. If IPR are financed with public money, it is necessary to calculate the appropriate price for them – if indeed it is appropriate to set a price. For example, should public university research be made available to the general public or merely to an exclusive circle?

4.4.2 Taxation and location

Taxation and location also raise problematic issues – for example, should the financial benefits of IPR transactions be taxed as capital gains? At a national level, the collection of such revenue must be secured, but there are still no clear and commonly accepted accounting rules on the treatment of IPR assets on company balance sheets. A similar challenge arises in respect of the physical location of the marketplace. As a wide circle of service providers is likely to emerge around the market, the host nation can expect additional benefits in terms of job creation and tax revenue.

4.4.3 IPR quality and measurement

IPR vary in quality according to the legislation on which they are based, as granting and application procedures differ in scope and stringency. The granting process has an important signalling function, while scope determines usability on the product markets. These two factors further influence enforceability in the event of a legal dispute.

Common rules and regulations to facilitate trading must therefore be developed, possibly on a global scale.

Additional uncertainty stems from the quantification and measurement of IPR. Without commonly accepted valuation principles, the commercial success of a given technology is determined by factors such as customer preference, industry investment and disruptive innovation, making meaningful and objective predictions difficult. Most methods of measuring these developments are at least partly subjective; only established technologies show measurable cash flow and therefore allow for commonly accepted valuations.

4.5 Financial products related to IPR markets

Financial products connect the IPR financial market with the IPR asset market as described in Section 2, describing the market model. Furthermore, their characteristics influence the trade in assets, i.e. patents. As such, we can differentiate between private vehicles, public–private vehicles and commoditisation vehicles.

4.5.1 Private vehicles

Private vehicles are financial instruments funded by investors from the private sector (as opposed to the public sector), and can be subdivided into two different instrument types: equity-based instruments (e.g. patent funds and their shares) and debt-based instruments (e.g. IPR securitisation and related financial products).

Equity instruments represent a residual interest in the issuer's net asset. This can, for example, be an ordinary share of company common stock, or a claim on the assets of and the return from a fund. The equity-based vehicle in the indicative market structure is shown in Figure 4.3.

The grey-shaded actors are directly involved in the equity-based vehicle. The vehicle is located at the crossroads of the IPR financial market and the IPR asset market, and thus assumes a connecting role: it connects the financially motivated investors with the patent market. During this process, the investors only interact with the vehicle itself, whereas the patent vehicle then buys, sells, licenses, etc. patents with the money provided by the investors with the goal of providing them with a promised return.

The equity-based vehicle may, for example, be structured as a company dealing with patents. This means that investors hold common

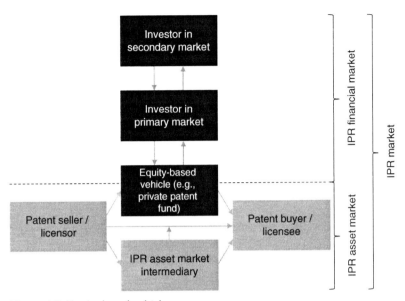

Figure 4.3 Equity-based vehicle structure.

stock in that company. The value of the shares is then defined by the investors' expectations regarding the company's performance. Investors can sell their shares to other investors on a secondary market.

Debt-based instruments (or liability instruments) are financial instruments that can require the issuer to deliver a financial asset to the owner of the instrument. Examples are products like bonds, or asset-backed securitisation. The debt-based vehicle in the indicative market structure is illustrated in Figure 4.4.

Similar to the equity-based vehicle structure, the grey-shaded boxes represent actors involved in a debt-based vehicle. Investors invest in the debt-based vehicle, and receive a return for that investment. In contrast to equity-based vehicles, investors in debt-based vehicles do not own part (share) of the vehicle, but are rather a liability. On the IPR asset market, the debt-based vehicle interacts with the patent market participants. In addition to intermediaries or the direct transfer of IPR, the vehicle buys/sells/licenses patents in order to provide the investors with the revenue streams agreed upon at the beginning of the contract. An IPR securitisation product may be similar to a bond structure: investors provide money, receive continuous payments to compensate for the risks taken and, in the end, receive their principal investment back.

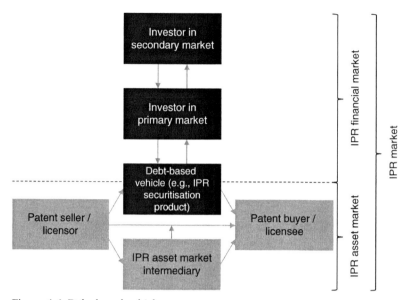

Figure 4.4 Debt-based vehicle structure.

4.5.2 Public–private vehicles

Public–private vehicles are financial instruments fully or partially funded by public money. They may be structured as a financial product suitable for trading on an IPR financial market. However, current examples are still limited to initiatives predominantly financed through the public sector. Public–private vehicles can be displayed in the indicative market structure as shown in Figure 4.5.

Although the aims pursued might be value adding (e.g. enabling national entities to exploit their patents effectively and provide advantages in terms of technology and knowledge), efficiency arguments have to be considered due to the funds' potential non-profit-orientated structure and public sector influence.

4.5.3 Commoditisation vehicle

The commoditisation vehicle aims at the continuous trading of the asset and is therefore suitable for licence rights rather than patents, as the latter are unlikely to be traded continuously. The IPX International serves as a role model for commoditisation vehicles'

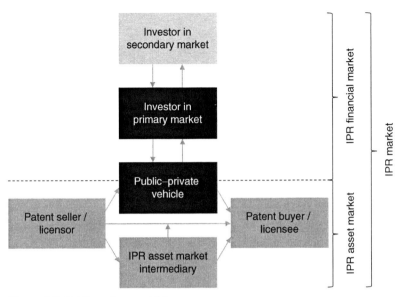

Figure 4.5 Public–private vehicle structure.

continuous trading exchange model. The commoditisation vehicle in the indicative market structure is shown in Figure 4.6.

The commoditisation vehicle's main achievement is its potential to enhance liquidity in the IPR asset market by making it similar to a financial market. Advantages associated with these characteristics include the attraction of a wider buyer and seller audience and the potential for more price transparency through continuous trading. However, a drawback associated with a more liquid IPR asset market is the potential for higher volatility since continuous trading is expected to set prices more often.

Table 4.1 provides an overview of tradeable financial products for an organised IPR market and depicts examples thereof.

While equity-based and commoditisation vehicles may invest in both early-stage and mature technologies (depending on the risk–return profile of the targeted investors), debt-based vehicles tend to be more appropriate for technologies that already generate stable revenues. This is due to the nature of debt-based instuments which usually reproduce recurring cash flows. On the other side of the spectrum, public–private vehicles may be used as a tool to foster early-stage technologies since they are backed by governmental institutions and are therefore at least partly aligned with public goals.

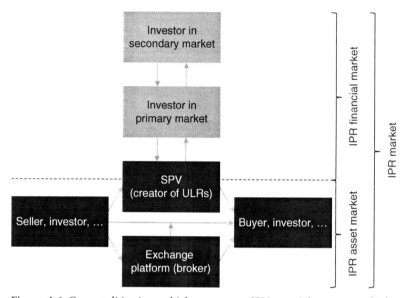

Figure 4.6 Commoditisation vehicle structure. SPV, special purpose vehicle; ULRs, unit licence rights.

In addition to the products in Table 4.1, further financial products (e.g. insurance products or classic loans) may be created. However, they do not have potential to be traded per se. Furthermore, index products based on patent strength of companies are not considered in our analysis since these represent an investment strategy rather than a direct involvement of IPR in the transactions. Looking forward, derivative products may also be considered once the IPR market has evolved further and IPR have become an accepted asset class.

4.6 Lessons learned

In moving towards a more formal IPR market in Europe, we highlight several areas of focus which should be considered by policy-makers, market organisers and companies that supply or acquire IPR.

Policy-makers and market organisers should consider focusing on the improvement of the IPR asset market before establishing an IPR financial market in order to avoid overheating of a potential marketplace. Furthermore, market fragmentation needs to be avoided and liquidity needs to be ensured to make the marketplace successful.

Table 4.1 *Overview of tradeable financial products*

Vehicle	Examples	Main funding market level	Technology suitability (early-stage or mature)
Private			
Equity-based	Patent funds (Patent Select I & II and Patent Invest 1, Germany)	IPR financial market	Early-stage and mature
Debt-based	Patent securitisation (Yale University, United States)	IPR financial market	Mature
Public–private	Public funds (LSIP, Japan and Intellectual Property Bank, Taiwan)	IPR financial market	Early-stage and mature (especially early-stage)
Commoditisation	IPXI's Unit License Rights	IPR asset market and IPR financial market	Early-stage and mature

A further objective of policy-makers can and should be ensuring high patent quality to support the growth of investor confidence. Further potential for market development lies in the advantages of licensing as opposed to transfer of ownership, leading to potentially higher homogeneity of IPR assets.

Companies that supply or acquire IPR should consider the opportunities presented by additional transaction partners in trading IPR with vehicles or financial products. These opportunities will develop even more with the emergence of commoditisation vehicles. Strategically, companies need to consider the development of the IPR market and its increasingly global scale, in order to consolidate or expand their competitive position. The market can even create new financing options for innovation, created by the emergence of the organised IPR market. In conclusion, the evolution of an IPR market may need companies to adopt a new view on their production process that integrates – rather than avoids – IPR as input factors.

References

Amihud, Y., & Mendelson, H. (1991). Liquidity, asset prices and financial policy. *Financial Analysts Journal*, 47, 56–66.

Arora, A. & Fosfuri, A. (2003). Licensing the market for technology. *Journal of Economic Behavior and Organization*, 52, 277–295.

Bader, M. A., Beckenbauer, A., Gassmann, O., König, T., Lohwasser, E., & Menninger, J. (2008). *One Valuation Fits All? How Europe's Most Innovative Companies Valuate Technologies and Patents*. Munich: PricewaterhouseCoopers.

Bader, M., Gassmann, O., Heilemann, U., Jha, P., Liegler, F., Maicher, L., Posselt, T., Preissler, S., Rüther, F., Tonnison, L., & Wabra, S. (2012). Creating an organised IP rights market in Europe. *Intellectual Asset Management*, 26, 33–38.

Becker, B., & Gassmann, O. (2006). Corporate incubators: industrial R&D and what universities can learn from them. *Journal of Technology Transfer*, 31, 469–483.

Benston, G. J., & Hagerman, R. L. (1974). Determinants of bid-asked spreads in the over-the-counter market. *Journal of Financial Economics*, 1, 353–364.

Black, F. (1971). Toward a fully automated stock exchange. *Financial Analysts Journal*, 27, 28–35+44.

Chordia, T., Roll, R., & Subrahmanyam, A. (2000). Commonality in liquidity. *Journal of Financial Economics*, 56, 3–28.

Chordia, T., Sarkar, A., & Subrahmanyam, A. (2005). An empirical analysis of stock and bond market liquidity. *Review of Financial Studies*, 18, 85–129.

Coase, R. H. (1960). The problem of social cost. *Journal of Law and Economics*, 3, 1–44.

Dahlman, C. J. (1979). The problem of externality. *Journal of Law and Economics*, 22, 141–162.

De Haan, J., Oosterloo, S., & Schoenmaker, D. (2009). *European Financial Markets and Institutions*. Cambridge: Cambridge University Press.

Fabozzi, F. J., & Modigliani, F. (1996). *Capital Markets: Institutions and Instruments*, 2nd edn. Upper Saddle River, NJ: Prentice-Hall.

Fabozzi, F. J., Modigliani, F., & Ferri, M. G. (1994). *Foundations of Financial Markets and Institutions*. Englewood Cliffs, NJ: Prentice-Hall.

Gambardella, A., Giuri, P., & Luzzi, A. (2007). The market for patents in Europe. *Research Policy*, 36, 1163–1183.

Gans, J. S., & Stern, S. (2010). Is there a market for ideas? *Industrial and Corporate Change*, 19, 805–837.

Gassmann, O., & Bader, M. A. (2011). *Patentmanagement: Innovationen erfolgreich nutzen und schützen*, 3rd edn. Berlin: Springer-Verlag.

Gassmann, O., & Becker, B. (2006a). Gaining leverage effects from knowledge modes with corporate incubators. *R&D Management*, 36, 1–16.

(2006b). Towards a resource-based view of corporate incubators. *International Journal of Innovation Management*, 10, 19–45.

Gassmann, O., Enkel, E., & Chesbrough, H. (2010). The future of open innovation. *R&D Management*, 40, 213–221.

Guellec, D., & Pottelsberghe, B. (2007). *The Economics of the European Patent System: IP Policy for Innovation and Competition*. Oxford: Oxford University Press.

Lichtenthaler, U. (2011). The evolution of technology licensing management: identifying five strategic approaches. *R&D Management*, 41, 173–189.

Lippman, S. A., & McCall, J. J. (1986). An operational measure of liquidity. *American Economic Review*, 76, 43–55.

Monk, A. H. B. (2009). The emerging market for intellectual property: drivers, restrainers, and implications. *Journal of Economic Geography*, 9, 469–491.

O'Hara, M. (1995). *Market Microstructure Theory*. Cambridge, MA: Blackwell.

(2003). Presidential address: Liquidity and price discovery. *Journal of Finance*, 58, 1335–1354.

Pagano, M., & Röell, A. (1996). Transparency and liquidity: a comparison of auction and dealer markets with informed trading. *Journal of Finance*, 51, 579–611.

Rivette, K., & Kline, D. (2000). *Rembrandts in the Attic: Unlocking the Hidden Value of Patents*. Boston, MA: Harvard Business School Press.

Roth, A. E. (2008). What have we learned from Market Design? *Economic Journal*, 527, 285–310.

Saunders, A., & Cornett, M. M. (2009). *Financial Markets and Institutions*, 4th edn. Boston, MA: McGraw-Hill Irwin.

Shapiro, C. (2001). Navigating the patent thicket: cross licenses, patent pools, and standard setting. In Jaffe, A., Lerner, J., & Stern, S. (eds.) *Innovation Policy and Economic Performance*, vol. 1, pp. 119–150. Washington, DC: National Bureau of Economics.

Shen, P. (2009). Developing a liquid market for inflation-indexed government securities: lessons from earlier experiences. *Economic Review*, 94, 89–113.

Smith, G., & Parr, R. (2000). *Valuation of Intellectual Property and Intangible Assets*. New York: John Wiley.

Tietze, F. (2011). *Managing Technology Market Transactions: Can Auctions Facilitate Innovation?* Cheltenham, UK: Edgar Elgar.

van Zeebroeck, N., van Pottelsberghe de la Potterie, B., & Guellec, D. (2009). Claiming more: the increased voluminosity of patent applications and its determinants. *Research Policy*, **38**, 1006–1020.

Williamson, O. E. (1985). *The Economic Institutions of Capitalism: Firms, Markets, Relational Contracting*. New York: Free Press.

5 | Valuation and rating methods for patents and patent portfolios

MARC BAUDRY

In the absence of a well-developed market for patents with numerous transactions, assessing the value of a patent essentially relies on private information retained by the patent holder. Even if such a market existed, the value of a patent would remain elusive, as it is highly dependent on the opportunities it offers, and those remain uncertain. In practice, professionals rely on a case-by-case thorough examination of patents to identify their perspectives of technological and economic success. Yet the strategies underlying this type of value assessment are disparate. They are generally classified in three groups.[1] The first group focuses on costs incurred to obtain patents. The implicit idea is that the total amount a rational inventor accepts to invest in order to obtain a patentable invention does not exceed the expected discounted returns from that invention. The second group intends to determine directly the expected cash flows generated by the patented invention. Finally the third group tries to identify comparable inventions in the same technological field and proceeds by analogy. In any case, the value assessment implies a lengthy and costly process. Thus, there is a clear need for valuation and rating methods that help economic agents to discriminate efficiently among a large set of patents and to detect early the more valuable patents. This chapter aims at presenting the state of the art. For such methods to be widely accepted by investors and patenting firms and to be used at least as a first screening tool before a more in depth and detailed examination, three fundamental requirements have to be fulfilled: transparency, objectivity and repeatability. Transparency implies that the theoretical foundations of the method are made explicit and that results of tests are disclosed so that any person skilled enough can check the overall quality and accuracy of the method. Transparency also implies that all the information used to

---

[1] See Goddar and Moser (2011) for a more detailed presentation of methods currently used by professionals.

assess the value of patents or to rate them is public or easily accessible. Objectivity means that the assessment or rating outcome does not depend on who implements it. For this purpose, characteristics of patents that are considered to implement the scoring method have to be as objective as possible. These two first requirements are limited de facto to assessment methods based on patent metrics. Indeed, patent metrics are either characteristics of patents directly obtained or computed from documents that have to be displayed for a patent to be granted or characteristics reported by the patent office. Repeatability essentially suggests that scores or values can be updated at regular and frequent intervals at a competitive cost. Econometric methods derived from published research constitute natural candidates that intrinsically meet the first two criteria and, most often, the third one also.

The econometric literature proposes indirect assessment methods based on observable and objectively measurable characteristics of patents, referred to as patent metrics. These metrics are assumed to determine the rent that may be extracted from patents. Assessment methods are said to be indirect in the sense that the level of the rent is not observed but has to be approached either by interview of patent holders or by observation of economic results, values or behaviours that are theoretically linked to this level. From this perspective, the evaluation process of a patent or a patent portfolio is close to the evaluation process of non-market goods, more specifically environmental goods. The objective of this chapter is to outline the various methods that can be implemented. It briefly presents the theoretical and econometric grounds of the methods while stressing whether they apply to an isolated patent or to a portfolio of patents, whether they lead to a ranking of patents or to an estimate of their monetary value and, last but not least, what they do or do not include in the rank assigned to a patent or in its estimated value. The pro and cons of each method are discussed, both from a theoretical point of view and from an empirical point of view.

Our review of the different econometric methods starts with a presentation of stated value approaches in Section 1. The chapter proceeds with indirect market value approaches presented in Section 2. Revealed value approaches are finally introduced in Section 3, and Section 4 proposes a synthesis. Such a typology and the corresponding denominations outline a parallel with valuation methods extensively used for non-market goods, more especially in environmental economics. In this spirit, stated value approaches are tightly linked to contingent valuation methods whereas indirect market value approaches are close to hedonic valuation methods in environmental economics. Though they do not have direct counterparts in environmental economics, revealed value approaches for patents exploit observed behaviour (mainly patent renewal decisions) of patent holders to infer the value of their patents, as the travel cost method does, for instance, to infer the private value of natural sites from the decision of households to visit them. The parallel with environmental economics helps to identify some common pro and cons but we recommend not to abuse it. Indeed, there are some key differences between patents and environmental goods. A first and trivial difference is that patents are policy instruments developed to solve some problems of externalities in innovation economics and are private assets, whereas environmental goods are natural assets subject to externality problems. The rationale for assessing the value of patents thus departs from the logic of environmental valuation.

5.1 Stated value approaches

5.1.1 General background

Stated value approaches, as we denote them, are probably the most intuitive of all types of approaches presented in this chapter for assessing patents' value. Their general strategy is to collect information directly from inventors or managers of innovative firms that hold patents. They thus avoid, or at least reduce to its minimum, the information asymmetry problem. Conversely, they are probably the most costly of all assessment methods that are discussed in this chapter. This may be detrimental given that the underlying econometric model has to be updated frequently to provide a relevant assessment tool for recent patents. The cost mostly comes from the extensive survey that requires contacting inventors or patent holders in order to ask them, in a form that will be discussed below, what is the private value of a given patent. The estimated econometric model then attempts to explain their answer by patent metrics. In essence and in terms of the associated econometric techniques, stated value assessment methods are close to contingent valuation methods for environmental goods. Both rely on statements of private agents about the value they associate with a good for which there is no well-developed market. Unsurprisingly, they share some advantages but also some drawbacks. If the comparison with contingent valuation methods in environmental economics helps to identify some key points that have to be discussed, other features of stated value methods for patents are specific to this domain.

At least three articles clearly apply a stated value assessment method. The first is an article by Harhoff, Scherer and Vopel (2003) which deals with patents grants with a 1977 German priority date. A similar methodology is implemented by Gambardella, Harhoff and Verspagen (2008) on data from a large-scale survey of European inventors that have been granted a patent with priority date 1993–1997 by the European Patent Office. The measure of value obtained for each patent in these two articles is given by the answer to the following question: 'Suppose that on the day on which this patent was granted, the applicant had all the information about the value of the patent that is available today. In case a potential competitor of the applicant was interested in buying the patent, what would be the minimum price (in Euro) the applicant should demand?' A slightly different question has been introduced in a similar study by Suzuki (2011) on Japanese patents but the method is essentially the same. The formulation of the question in the three aforementioned articles deserves several comments that may also concern contingent valuation methods in environmental economics. First, as mentioned by some of these authors, what is measured is a reserve price of the patent holder for the patent, not a market value of the patent. Indeed, it might be the case that no buyers could be found at this value so that the transaction would never occur. This is in line with the concept of willingness to receive in contingent valuation methods for environmental goods which is a measure of the private value of an environmental good, not of its market value. Second, because private agents are facing a hypothetical transaction, they may lack some benchmark to declare a value. To remedy this problem, the articles by Harhoff, Scherer and Vopel (2003) and Gambardella, Harhoff and Verspagen (2008) follow a common practice in contingent valuation methods and offer a menu of ten intervals for responses.[2] Such a menu induces a lack of accuracy of the values in the responses. Nevertheless, it is supposed to be counterbalanced in

---

[2] These intervals are less than €30K, between €30K and €100K, between €100K and €300K, between €300K and €1M, between €1M and €3M, between €3M and €10M, between €10M and €30M, between €30M and €100M, between €100M and €300M, and more than €300M.

terms of realism of responded values as long as the intervals are fixed on the basis of previous works on patent values that yield an idea of the general shape of their distribution.[3] Suzuki (2011) avoids having to define a scale in monetary value. Indeed, in the survey used by Suzuki (2011), respondents position their patent in the whole population of patents by indicating to what quartile of the distribution of values over the whole population of patents they think their patent belongs. Third, due to the hypothetical choice they are facing, respondents may intentionally or unintentionally give biased responses. Gambardella, Harhoff and Verspagen (2008) address this problem by comparing answers from inventors and from managers for a subset of patents. They find that inventors slightly overestimate the value of their patent, but the bias is negligible. However, they consider only one possible type of bias. For instance, their result is not inconsistent with both inventors and managers being intrinsically reluctant to acknowledge they have low-value patents even if they have guarantees that answers will be treated anonymously. So far, comments on stated value methods have been confined to aspects that are not specific to an application to patents, except to some extent the last aspect.[4] We now turn to a more specific aspect, the exact content and interpretation of declared values.

5.1.2 *What do stated values encompass?*

Harhoff, Scherer and Vopel (2003) and Gambardella, Harhoff and Verspagen (2008) argue that the method they present and implement in their article is consistent with a broad definition of the value of a patent. They more specifically stress that they measure the 'asset value' of patents rather than the narrower 'renewal value' of patents. The 'renewal value' refers to what is lost if the renewal fee is not paid and the patent falls into the public domain. The patent holder then loses all net revenues that accrue from the exclusivity right on industrial and commercial opportunities that the patent confers. The 'asset value' adds to the 'renewal value' the potential loss due to the new owner blocking revenues from complementary inventions. Stated in another way, the 'asset value' is associated with a transfer of the right whereas

---

[3] With this respect, we have a first argument to consider that the different valuation methods are complements rather than substitutes.
[4] But the problem of biases is a well-known problem of contingent valuation methods in environmental economics.

the 'renewal value' is associated with an end of the right. Note that it is possible to modify the question used in the survey to measure a 'renewal value' rather than an 'asset value'. For instance, rewriting the question as follows would provide a 'renewal value': 'Suppose that on the day on which this patent was granted, the applicant had all the information about the value of the patent that is available today. In case a front fee (in place of renewal fees effectively paid) has had to be paid at the date of grant for holding the patent the same length of time it has effectively been held, what would be the maximum front fee (in Euro) the applicant should accept?' It is expected that the answered value would be lower than with the initial question.

A more important point that is not discussed by the authors is whether their method correctly captures the option value of patents or not. The fact is that, as the question is formulated, the method is probably not able to capture the option value of patents. In order to make this point more explicit, let us consider two patents A and B with the same initial rent at the date of grant and the same stochastic process that governs the dynamics of the rent. Let us consider for instance a geometric Brownian motion to describe the dynamics of the rent with an additional Poisson process that can make the rent suddenly fall to zero. Assume that each patent effectively generates the corresponding rent if and only if the patent holder decides to invest in the manufacturing and commercialisation of the patented invention. Due to sunk investment costs and to the uncertainty surrounding the dynamics of the rent, the decision to invest has to be analysed as a real option. Assume that the initial rent justifies renewing the two patents to keep the option to invest alive but is not high enough to decide to invest immediately. Then, by construction, *Ex ante* (i.e. at the date of grant) the two patents have the same option value. Nevertheless, assume that in case A the time path of the rent justifies exercising the option to invest ten years after the grant whereas a jump down to zero occurs at age five for the rent in case B and the option to invest will therefore never have been exercised. *Ex post* (i.e. at least after age five), A has a positive value whereas B is worth zero. What is expected with the question initially formulated is that respondents give the *ex post* value, not the *ex ante* value. Stated in another way, it is expected that they answer as if they had perfect foresight of the time path of the rent. Note that this is stronger than the usual assumption of rational expectations. Indeed rational expectation implies that economic agents exploit all the information available at the date of grant whereas perfect foresight is built on the unrealistic assumption that they also have perfect information about what will be realised in the future. Clearly, what is needed for patent scoring is an *ex ante* valuation. Note that the aim of the three articles discussed in this section was not initially to develop a scoring method but rather to get some insights into the distribution of patents value as effectively realised and to explore some links with patent metrics. With this aim in view, measuring *ex post* values makes sense. What we point out in this discussion is that the method would have to be adapted in order to use it as a basis for patent scoring. A first step towards this goal could be to modify the initial question as follows: 'Suppose that on the day on which this patent was granted, a potential competitor of the applicant had been interested in buying the patent. What would have been the minimum price (in Euro) the applicant should have demanded according to the information available at that date?' The date of grant can be interchanged with any other date between the date of grant and the current date. Whatever the date considered, the answer is theoretically the estimated value under rational expectations. Nevertheless, it is doubtful that a respondent could disregard information obtained between the date which is refered to in the question and the current date if the two dates differ. Therefore, a precaution probably consists in asking respondents to give the value at the current date and to consider patents of different ages in the survey.

5.1.3 Econometric model

Whatever the exact form of the question, the econometric treatment of the answer is that of an ordered discrete choice model.[5] Some technical details help in understanding how to use the estimated model for rating patents. Consider that the value of a patent results from observed and unobserved factors that act in a multiplicative manner. Accordingly, the value V_i of a patent i may be written as

$$V_i = f\left(x_i^1, \ldots, x_i^K\right)\varepsilon_i \tag{5.1}$$

where f is a function of the measures x_i^k ($k \in \{1, \ldots, K\}$) of K metrics (i.e. characteristics either directly reported in official documents published in counterpart of the grant or reported by the patent office) obtained

---

[5] We only consider the case of a menu of interval responses.

for patent *i*. The function *f* thus captures observed heterogeneity between patents. Conversely ε_i is a random term that captures unobserved heterogeneity between patents; ε_i is assumed to be identically and independently distributed across patents. The multiplicative form of (5.1) is more specifically suitable to ensure that V_i takes only positive values. This is guaranteed, for instance, if a Cobb–Douglas form is used for *f* and ε_i is assumed to have a log-normal probability distribution. Let Ψ be the cumulative density function of the random term in (5.1) and let v_m ($m \in \{1, \ldots, M\}$) be the generic term to denote the bounds separating the interval responses. Then, the probability of obtaining an answer that belongs respectively to the first ($m = 1$), an intermediate ($m \in \{2, \ldots, M - 1\}$) or the last interval ($m = M$) is

$$\Pr[V_i \leq v_1] = \Psi\left(v_1/f\left(x_i^1, \ldots, x_i^K\right)\right) \tag{5.2a}$$

$$\Pr[v_m < V_i \leq v_{m+1}] = \Psi\left(v_{m+1}/f\left(x_i^1, \ldots, x_i^K\right)\right) \\ - \Psi\left(v_m/f\left(x_i^1, \ldots, x_i^K\right)\right) \tag{5.2b}$$

$$\Pr[V_i > v_M] = 1 - \Psi\left(v_M/f\left(x_i^1, \ldots, x_i^K\right)\right) \tag{5.2c}$$

These probabilities are directly used to write the log-likelihood of observed responses and then estimate the unknown parameters in the functional form *f* by a standard log-likelihood maximisation procedure.[6] V_i is the latent variable of the ordered discrete choice model and is estimated by

$$\hat{V}_i = f\left(x_i^1, \ldots, x_i^K\right) \mathsf{E}[\varepsilon_i] \tag{5.3}$$

where E stands for mathematical expectation. Equation (5.3) is valid both for patents used for estimation purposes and for other patents, and can thus serve as a basis for ranking patents according to their expected monetary value. Note that whereas bounds of intervals are exogenous in Harhoff, Scherer and Vopel (2003) and Gambardella, Harhoff and Verspagen (2008), the different form of the question used in the survey implies that they have to be estimated along with the unknown parameters of the functional form *f* in Suzuki (2011).

---

[6] See for instance Greene (1993), Chapter 19. If the random terms are log-normally distributed, (5.2a) to (5.2c) may be equivalently expressed in natural logarithms so that the resulting model is a standard ordered probit model.

5.1.4 Empirical results and perspectives

The metrics used by Harhoff, Scherer and Vopel (2003) as explanatory variables are the scope of patents measured on the basis of International Patent Classification (IPC) classification codes, forward citations, backward citations, references to non-patent literature, the family size of patents and outcome of opposition cases.[7] The metrics introduced by Gambardella, Harhoff and Verspagen (2008) are dummy variables for the number of forward citations, backward citations, claims, and the number of designated countries in which the patent is applied for. Finally Suzuki (2011) considers forward citations received within five years of patent acquisition, references to scientific publications and a dummy for the appurtenance to a triadic family. Table 5.1 displays some details on the results obtained in these three articles and compares them with results obtained by other authors with the other valuation methods. Note that the logic of the model is that patent metrics explain the monetary value of patents. However, one can wonder if the family size and forward citations are not an outcome of the economic value of patents. This point is worth noting given that Harhoff *et al.* (1999) previously used similar data to estimate an inversed relation. We argue in favour of a causality effect from forward citations to economic value. Indeed, forward citations are obtained by a patent because of its technical importance. Technical importance, in turn, is likely to generate economic revenues but these revenues are increased by the existence of forward citations due to a positive and credible signal that helps economic partners of the patent holder (e.g. banks, stakeholders, other firms that cooperate in the R&D programme) to identify the patent as a patent of high technical importance and the patent inventor as a productive and key inventor. Broadly speaking, in all three articles it was found that patent metrics have significant individual or group impacts on the latent variable but that they only explain a small fraction of the total variance of patent value. Gambardella, Harhoff and Verspagen (2008), for instance, obtain that their model explains at best 11.3% of the variance of patent value and conclude that 'the measure of our ignorance is still sizable'. Therefore,

[7] The expected impact of the different patent metrics has been extensively discussed in the literature. See, among others, Omland (2011) for a comprehensive review of these impacts.

Table 5.1 *Patent metrics and patent characteristics used in studies mentioned*

	Stated value			Indirect market value			Renewal method		
	Harhoff et al. 2003	Gambardella et al. 2008	Suzuki 2011	Bloom & Van Reenen 2002	Lanjouw & Schankerman 2004	Hall et al. 2005	Barney 2002	Bessen 2008	Baudry & Dumont 2012
Forward citations	***	***	***	0	***	***	?	***	***
Backward citations	***	**			***			***	0
References to non-patent literature	***		***						
Claims		***	0		***		?	***	0
Technological scope	*								**
Number of designated countries	***	***							0
Patent Cooperative Treaty (PCT) filing strategy									0
Family size	***								0
Triadic family			***						
Number of priority							?		0

Existence of licensing	***	**
Outcome of opposition process	***	
Outcome of annulment process	***	
Number of words per independent claim	?	
Number of keywords	?	
Length of written specification		**

*** statistically significant at a 1% confidence level;
** statistically significant at a 5% confidence level;
* statistically significant at a 10% confidence level;
0 statistically not significant;
? significance not reported.

it is doubtful that a forecast of the value of new patents based on the estimated coefficients of the model and the observed metrics of these new patents helps in efficiently discriminating between them.

The three articles discussed in this section intend to measure value at the patent level, not at the level of the portfolio. Nevertheless, it is often argued that the portfolio level is more relevant than the single patent level when trying to assess patents. The underlying idea is twofold. Firstly, in complex technologies multiple complementary patents are required to develop a new product or production process. In this case, a single patent is worthless and what may be subject to a transaction is only the group of complementary patents. Secondly, patent holders may adopt a patenting strategy that aims at developing patent fences and blocking competitors by applying multiple patents protecting close inventions. Acquiring a single patent is then worthless because the risk of infringement of the other patents is high. Asked about the reserve price for a patent that is tightly linked to other patents for one of these two reasons, the patent holder will try to determine the loss of value for his portfolio of patents incurred if the patent is transferred. Due to the existence of technical or strategic complementarities, the declared value will then be higher than the value of the patent taken alone but lower than the value of the whole portfolio, except in the case of perfect complementarity. In a more recent study, Gambardella, Harhoff and Verspagen (2013) address this problem and suggest controlling for the size of the portfolio and adding a question about the value of the portfolio of patents linked to the patent concerned. The way the question is worded leaves the identification of the relevant portfolio to the discretion of the patent holder. Gambardella, Harhoff and Verspagen (2013) conclude in favour of the existence of complementarities and confirm that focusing on the value of isolated patents may be misleading.

5.2 Indirect market value approaches

5.2.1 General background

Some goods or assets do not need to be exchanged on a dedicated market for having an implicit market value. This is typical of goods or assets which value is capitalised in the market value of other goods or other assets. For instance, environmental characteristics and local public goods are non-market goods but their private value is capitalised in the price of houses. Such a principle has been extensively used in the hedonic valuation methods in environmental economics. Again, and even if the link has never been stressed to our knowledge, a class of patent valuation methods also exploits this principle. In papers that belong to this hedonic tradition, the stock market is considered as a market that capitalises the value of both tangible and intangible assets of a firm. The general idea is that firms choose their flow of investments in tangible and intangible assets in a way that maximises their expected and discounted sum of present and future benefits. Substituting the optimal investment choices in the expression of the expected and discounted sum of benefits yields the maximum value an external investor should accept to pay for acquiring the firm. Such a value theoretically depends on the current stocks of tangible and intangible assets and is referred to by some authors (see, for instance, Morricone and Oriani (2011) or Bessen (2009)) as the hedonic value of the firm. Tangible assets are measured by their book value but there is no direct measure of intangible assets. Following Griliches (1981), a broad strand of empirical literature has tested different proxy measures of intangible assets. More specifically, measures based on R&D expenditure, patent counts, patents granted per unit of R&D expenditure, forward citation counts, forward citations received per patent, either alone or combined, have been used. These empirical studies estimate the hedonic value function of firms and then infer the implicit value of patents or patent characteristics. They initially deal with the whole portfolio of patents of each firm. Nevertheless, because they focus on counts or averages rather than on more complex indicators related to the structure of portfolios (e.g. indicators of the proximity of patents), most of these studies are not able to deal correctly with the technological or strategic complementarities that make the value of a portfolio higher than the sum of its components. Moreover, their concept of portfolio refers to the whole set of patents held by the same firm not to subsets of patents that are tightly linked. Roughly speaking, for the time being they assess the value of the whole set of patents a firm holds by assuming it is just the sum of isolated patents. The rest of this section describes and comments on the common aspects of the estimated models, it then presents the results obtained as regards the impact of patent metrics and it finally addresses a fundamental question as regards its use for patent rating.

5.2.2 Econometric model

Since Hayashi (1982), it is well known that in the presence of adjustment costs and under linear homogeneity of the profit function and the adjustment cost function, the inter-temporal maximisation of the value of a price-taking firm is consistent with Tobin's Q theory of investment. A key result for empirical studies is also that the value of a firm is proportional to its stock of capital and that the proportionality factor is Tobin's marginal Q which is identical to Tobin's average Q. Two extensions of the model are of interest for the valuation of patents. First, Hayashi (1982) himself shows that if a firm is price-maker, then the marginal Q is lower than the average Q. In the modified relation between the value of the firm and the stock of capital, nominal rents that accrue from market power have to be added to the stock of capital. This is of importance when dealing with a stock of knowledge protected by patents. Indeed, patents confer an exclusivity right on commercial and industrial opportunities associated with the patented invention and it is expected that this exclusivity right generates a rent. A second extension is due to Hayashi and Inoue (1991) who examine the case of multiple capital goods. They show that if the profit function is weakly separable with respect to capital inputs and if the capital aggregator is homogeneous of degree one, then a proportional link exists between the value of the firm and the capital aggregator. They more specifically suggest a linear approximation of the aggregator where the different capital goods are weighted by their user's cost of capital that just equals their marginal productivity at equilibrium. If we distinguish between tangible assets and knowledge capital, the functional relation between the different assets of a firm and its market value may thus be formulated as

$$W_{it} = q_{it}(A_{it} + \gamma K_{it} + R_{it}) \tag{5.4}$$

where W_{it} denotes the market value of firm i at date t, A_{it} denotes the book value of tangible assets, K_{it} denotes the firm's knowledge assets and R_{it} denotes the rent that accrues from patents granted to firm i. The parameter γ is the implicit or shadow relative price of knowledge assets in terms of tangible assets, and q_{it} is the price at which the package of tangible assets, intangible assets and rents is valued by stock markets; this price may be assumed to depend on unobserved factors ε_{it}. It is more specifically convenient to express this price as $q_{it} = \bar{q}_t \exp(\varepsilon_{it})$.

After some transformations and approximations, expression (5.4) may be rewritten as[8]

$$\ln W_{it} = \ln \overline{q}_t + \ln A_{it} + \gamma(K_{it}/A_{it}) + R_{it}/A_{it} + \varepsilon_{it} \qquad (5.5)$$

or equivalently

$$\ln Q_{it} = \ln \overline{q}_t + \gamma(K_{it}/A_{it}) + R_{it}/A_{it} + \varepsilon_{it} \qquad (5.6)$$

where $Q_{it} = W_{it}/A_{it}$ denotes Tobin's average Q for firm i at date t and the intercept of the model can be interpreted as an estimate of the logarithmic average of Tobin's Q for each year. Treating ε_{it} as independently and identically distributed random terms yields the econometric model to be estimated. Last but not least, proxy variables for the unobserved stock of knowledge and amount of rents have to be used. The stock of knowledge is the outcome of an innovation process with R&D expenditures as the main input. Therefore, the sum of past R&D expenditures weighted by a depreciation rate is generally used as a proxy for the stock of knowledge.[9] But, in order to reflect the heterogeneity of firms as regards their ability to produce valuable knowledge, patenting rates defined as the ratio of patent counts PC_{it} to R&D expenditures RD_{it} are introduced as a key element of the productivity of R&D expenditures. Similarly, rents R_{it} may be assumed to depend on patent counts PC_{it} eventually weighted by citations received (forward citations FC_{it}) to account for heterogeneity in their quality. Depending on the exact formulation used for the proxy variables, different specifications of (5.6) are obtained. They generally imply R&D expenditures, patent counts and citations counts as explanatory variables, either directly or in the form of

---

[8] The approximation x of $\ln(1 + x)$ for small values of x is more specifically used. Its main advantage is that the resulting econometric model is linear in variables and thus estimable with basic econometric methods. Hall, Jaffe and Trajtenberg (2005) argue that it is doubtful the ratio of intangible to tangible assets is small and reject this approximation. They thus have to estimate their model by non-linear least squares.

[9] The choice of the annual depreciation rate is somewhat arbitrary and varies from 15% in Hall, Jaffe and Trajtenberg (2005) to 30% in Bloom and Van Reenen (2002). In footnote 26, Hall, Jaffe and Trajtenberg (2005) note that 'Small departures from this rate do not make a difference to the results, but this is an issue that deserves some serious revisiting, in light of the much more detailed data that we now have at our disposal.' With this respect, it would be more satisfactory to compute stocks of renewed patents as the withdrawal of a patent reveals its obsolescence.

ratios. Note that if the model aims at estimating the implicit value of patent characteristics it is essential that both these characteristics and patent counts are introduced as covariates along with R&D expenditures. Indeed, in the absence of patent counts, it may be unclear whether patent characteristics (e.g. counts of forward citations) capture the productivity of R&D expenditures or the quality of patents granted. When the model controls for R&D expenditures, patent counts and eventually some indicators of the structure of the portfolio, estimates of the parameters can be used to compute the implicit value of a patent as the combination of the implicit values of its characteristics.

It is assumed in the neoclassical approach to Tobin's Q that economic agents have rational expectations of future flow of profits. The option value of future opportunities of private returns offered by patented invention should then logically be included in the market value of firms. This is suggested by Bloom and Van Reenen (2002) who interpret the positive difference between the value of patents estimated by a Tobin's Q method and the value inferred from the estimation of a production function (with real sales as the output) as an empirical evidence of the existence and importance of option value. Nevertheless, their model also suggests discriminating between patents protecting embodied knowledge and patents protecting disembodied patents. Knowledge is said to be embodied if the development of industrial and commercial applications of the patented invention has been undertaken. Embodied knowledge thus generates a rent from production and sales. Conversely, knowledge is said to be disembodied if industrial and commercial opportunities secured by patented inventions have not yet been undertaken. Nevertheless, disembodied patents may generate pecuniary advantages. Indeed, the grant of patents constitutes by itself a signal of the inventiveness of a firm and eases its access to external financing. Since the two types of patents do not generate the same kind of rent, the distinction between embodied and disembodied patents matters in the hedonic equation of the value of a firm. Measures of the stock of patents used in Tobin's Q methods disregard the distinction between the two types of patents. It thus appears that existing Tobin's Q studies aiming at estimating the value of patents do not correctly address the existence of an option value of patents and may therefore be subject to misspecification.

5.2.3 Results and perspectives

The oldest studies dealing with knowledge capital in Tobin's Q hedonic equations (Griliches 1981; Cockburn and Griliches 1988) consider basic patent counts as the sole proxy variable for knowledge capital. Therefore, they only provide estimates of the mean value of patents. As Bessen (2009) notes, if the distribution of patent values is highly skewed, some convergence problems may arise when estimating the hedonic equation. Indeed, the mean value is eventually infinite for some distributions such as the Pareto distribution which is found to better fit the upper tail by Silverberg and Verspagen (2007). Another important problem that arises in these studies and also features in more recent works is the construction of patent stocks. Firstly, existing studies generally focus on stocks of patents applied for and granted by a single patent office, most often the US Patent and Trademark Office (USPTO) or the European Patent Office (EPO). Thus, the stock of patents used for estimating the hedonic equation is only a fraction of patents that firms hold. In addition to an eventual misspecification problem, one may wonder what value exactly is assessed. If for instance USPTO patents have 'brother' patents granted by the EPO but not included in the measure of the stock, what is assessed is rather the value of the whole family of USPTO and EPO patents than the value of USPTO patents alone. Secondly, stocks of patents are constructed using a declining balance formula with a constant depreciation rate. The first drawback is that, in order to limit the effect of missing initial stock levels, stocks are computed over the whole period covered by the data but the estimation is conducted on a sub-period that excludes the beginning of the period. A loss of information results from this shortening of the initial data set. The second drawback is that the depreciation rate is fixed exogenously and somewhat arbitrarily. A way to limit this drawback could consist in computing for each date stocks of patents that are not yet withdrawn. To our knowledge and for the time being, this has not been implemented, probably because information on the status of patents is not available in large data sets used for Tobin's Q studies.

An important step towards the use of Tobin's Q methods for patent scoring has been crossed with the work by Bloom and Van Reenen (2002) and Hall, Jaffe and Trajtenberg (2005) who introduce patent forward citations in their respective empirical studies. Bloom and Van Reenen (2002) consider a citation-weighted patent stock whereas Hall, Jaffe and

Trajtenberg (2005) introduce patent stocks and citation stocks as separate regressors in the hedonic equation. They argue that it is 'a way of gauging the enormous heterogeneity in the value of patents'. These two studies find empirical evidence that citations contribute to improve significantly the quality of the regression. Lanjouw and Schankerman (2004) go further by considering a weighted sum of patents where the weights depend on an index of the overall quality of patents. This index is computed as the latent variable of a factor model with claims, backward citations and forward citations within five years of the patent application as the three observed indicators (family size was finally excluded because it was only available for about 20% of patents in the database). A key advantage of their model to proceed with patent scoring is that the heterogeneity of patents is captured by multiple metrics through the factor model. As a result, a patent may be ranked higher than another one because it exhibits far more claims and backward citations in spite of fewer forward citations. Stated another way, forward citations are not the only determinant of the rank of patents. Actually, the factor model from which the index of quality is inferred could suffice for ranking purposes. Its introduction as an explanatory variable in Tobin's Q equation just serves to produce a monetary value of firms' patent stocks.

Even if, from a purely econometrical point of view, Tobin's Q studies are suitable for producing monetary estimates of patents portfolios or isolated patents, one may wonder if, from a theoretical point of view, it does not pose some problems. Indeed, a key feature of any Tobin's Q empirical study is that the assumption of informational efficiency of stock markets is satisfied. A first and rather common criticism should of course consist in pointing out that empirical proofs of this assumption are highly controversial. A second criticism is more intrinsic to the implementation of Tobin's hedonic equation to produce patent scoring. Note that what is targeted by the criticism is neither Tobin's hedonic equation with patents or patent metrics as proxy variables for intangibles nor the principle of patent scoring but the use of Tobin's equation to score patents. Indeed, the goal of the aforementioned empirical studies on Tobin's hedonic equation with patents or patent metrics as explanatory variables was initially to address the problem of measuring the pace of innovation. More specifically, what was at stake was the measure of a quality or value index of patents to complement simple patent counts. With this aim in view these studies exploit the alleged informational efficiency of stock market to obtain such a measure. They thus implicitly assume or explicitly test *ex post* that investors rationally deal with publicly available patent metrics and account for these data to value a firm's patents. Commenting on their results, Lanjouw and Schankerman (2004) for instance assert that 'investors have enough information to distinguish differences in mean patent quality across companies'. Similarly, commenting on the difference they obtain between Tobin's Q valuation method and an estimate based on productivity gains, Bloom and van Reenen (2002) write that 'this larger point estimate on the current value obtained in the market value equation appears to reflect the forward looking nature of the market value measure'. In essence, no additional information compared to that available to financiers is thus produced by empirical studies of Tobin's hedonic equation. Resulting patent scoring cannot therefore be thought of as a tool that helps them better discriminate among patents. The same criticism applies to event studies on patents. The paper by Austin (1993) is the first to implement an event study on a class of biotechnology companies. Events are identified as the announcement in the *Wall Street Journal* of the grant of the patent. Though Austin (1993) shows that such events generate significant abnormal returns for patenting firms, only very few papers have followed this direction of research. A reason for this is that, in addition to the constraint that firms are listed on the stock market (a constraint that also applies to Tobin's Q methods), it is required that granted patents are 'solitary'. This means that these patents are issued in weeks when no other patent in the sample is issued, which is a strong requirement.[10] The study by Hall and Mac Garvie (2010) constitutes an exception but it attempts to assess the overall effect of court decisions as regards the patentability of software in the United States rather than the value of individual patents.

5.3 Revealed value approaches

5.3.1 General background

This last section finally introduces revealed value approaches. Valuation methods belonging to this category build on the fact that some decisions or results of a patenting firm objectively depend on revenues

---

[10] This is required to make sure potential abnormal returns are attributable to the patent being issued. Note that because an effect is expected on rivals, it is required that not only the firm which is granted the patent should not have other patents issuing in the week but rival firms also.

generated by its patents. In this sense, these methods are close to revealed preference methods used for environmental valuation. Their strength is to rely on observed facts and, as such, they are not subject to the type of biases that affect stated value assessment methods. Moreover the underlying economic assumptions are less subject to criticism than the assumption of informational efficiency characterising indirect market value approaches. Their weakness is that they often disregard some components of the total economic value of a patent or portfolio of patents.

A first method in this group seeks to estimate the gains generated by intangible assets, more specifically patents. For this purpose, real sales of a panel of firms are econometrically estimated as the output of a production function whose inputs are tangible assets, labour used and its stock of granted patents or a combination of metrics of patents granted as a proxy variable for its intangible assets. The method was introduced by Bloom and Van Reenen (2002) but, to our knowledge, it has never been implemented by other authors since. The method is better adapted to the valuation of a portfolio of patents than to an isolated patent, unless a sum of patent counts weighted by citations or a quality index is used to compute patent stocks. It is reminiscent of Tobin's Q method in the sense that the right-hand side of the estimated equation is quite similar but differs in the dependent variable on the left-hand side. A first advantage is that the method is applicable to all companies, listed or not. Educational institutions and research institutions cannot be included in the analysis, which may still introduce a selection bias in patents treated. A second advantage is that the dependent variable does not reflect the assessment of the value of patents held by external economic agents but is directly linked to economic results obtained by the patented firm. Nevertheless, the method only captures current or short-term productivity gains. Long-term and uncertain productivity gains are disregarded. Yet long-term productivity gains are generally considered as a major source of the option value of patents. This is probably a major reason explaining why the method has not spread to other empirical studies on patent values.

Another method that belongs to the group of revealed value approaches is the renewal decision approach. As long as patent holders adopt a rational value maximising behaviour to renew their patents, their decision to either pay the renewal fee in order to keep the patent alive or withdraw the patent indirectly reveals the value they attribute to this patent. As in stated value methods, the value revealed by the renewal or withdrawal decision is actually the reserve price of the patent holder. Moreover, in case of a patent embedded in a portfolio of patents that are linked due to technological or strategic complementarities, the reserve price refers to the loss in the value of the portfolio that would result from the withdrawal of the patent. The concept of value is thus closer to that in stated value methods than to that in Tobin's Q methods. Our presentation follows the main lines of Baudry and Dumont (2012) and starts with a general option like formulation. We consider a patent applied for at date t and that generates an annual rent R_t^a at age a. The rent encompasses all pecuniary advantages that accrue from the retention of the exclusivity right on industrial and commercial applications of the invention secured by the patent. Whether the patent is embodied or disembodied does not theoretically matter. Nevertheless the distinction between embodied and disembodied patents would be required to explicitly model the switch from disembodied knowledge to embodied knowledge and is systematically neglected in existing studies. What is considered in the option model is thus the opportunity to postpone the withdrawal of the patent. As Baudry and Dumont (2006) note, the corresponding option model may be interpreted as a succession of nested European Puts with the rent as the underlying asset value, the renewal fee as the exercise price and a maturity that corresponds to the lapse of time between two payments of the renewal fee (one year in most European countries, four years in the United States). The option problem considered by Bloom and Van Reenen (2002) and discussed in the previous section devoted to Tobin's Q methods is different. What Bloom and Van Reenen (2002) model as an option is the opportunity to switch from a disembodied patent to an embodied patent. The corresponding option model may be interpreted as an American Call with the rent associated with disembodied and embodied patents as the two underlying assets, the sunk cost of disembodiment as the exercise price and the statutory lifespan as maturity.[11]

---

[11] Bloom and Van Reenen (2002) simplify their analysis by considering an infinite life of patents. They do not for their part consider the decision to pay the fee and renew the patent or withdraw the patent.

5.3.2 Econometric modelling of renewal decisions

The option problem considered in renewal models may be written as

$$
V_t^a = \text{Max} \begin{cases} R_t^a - f_t^a + \dfrac{E_{t+a}\left[V_t^{a+1}\right]}{1+r} & \text{if the patent is renewed} \\[2mm] 0 & \text{if the patent is withdrawn} \end{cases} \qquad \forall a < A
$$

(5.7a)

and

$$
V_t^A = \text{Max} \begin{cases} R_t^A - f_t^A & \text{if the patent is renewed} \\[2mm] 0 & \text{if the patent is withdrawn} \end{cases}
$$

(5.7b)

where V_t^a is the value of the patent at age a and A is the statutory life limit of patents. We denote by f_t^a the renewal fee that applies at age a for patents applied for at date t. Renewing the patent yields the rent R_t^a net of the renewal fee f_t^a plus the expected and discounted value $E_{t+a}\left[V_t^{a+1}\right]/1+r$ of the patent at age $a + 1$ where r stands for the discount rate and E_{t+a} for the mathematical expectation conditional on all information available to the patent owner at date $t + a$. This last term vanishes if the statutory life limit A is reached because the patent will then fall into the public domain at the next period of time and the corresponding rent will thus dissipate. If the patent is not renewed, no renewal fee is paid but the rent immediately falls to zero due to the loss of the exclusivity right. It is worth noting that, following the terminology used by Gambardella, Harhoff and Verspagen (2008) and commented on in Section 1 of this chapter, the value function V_t^a reflects the 'renewal value' of patents but not their 'asset value'. Moreover, renewal methods provide an estimate of the value of isolated patents but are not adapted for the time being to portfolios of patents. Indeed, this would require assuming some correlation between initial rents and/or random shocks affecting the dynamics of rents in order to capture the dependence between patents of the portfolio. Ignoring such a dependence implies that the value of a portfolio is nothing other than the sum of values of the patents it includes.

Renewal fees play a crucial role for obtaining monetary values of patents as they are the sole observable cost of renewing patents. There exist other costs that may be much higher but are not observed. These encompass, among others, internal costs to assess the usefulness of the patent and enforcement costs. Unobserved costs are implicitly subtracted from the rent R_t^a that actually encompasses all unobserved pecuniary advantages and disadvantages of patents. Accordingly, the rent is treated as the state variable of the dynamic programming (5.7) and is assumed to be known to the patent holder but not to other economic agents. The asymmetry problem is thus fully handled by renewal models. This point dramatically contrasts with Tobin's Q methods. Nevertheless, important restrictions on the dynamics of the rent are generally imposed in the literature. A reason for this is that the use of stochastic processes to describe this dynamics yields complex exercise rules for the option problem. Pakes (1986) and the three subsequent studies by Lanjouw (1998), Baudry and Dumont (2006) and Deng (2011) are illustrative of this point. None of these authors consider patent metrics to introduce a source of heterogeneity across patents and their models thus provide only a distribution of patent values. Therefore, following the pioneering article by Schankerman and Pakes (1986), most of the econometric literature on patent renewal decisions imposes a simplifying restriction on the dynamics of the rent. The simplification consists in assuming that the rent decreases while renewal fees increase as the patent ages. The increase of renewal fees is an ad hoc assumption based on observed profiles of renewal fees in most patent offices. What is effectively captured by the decrease of the rent depends on whether the patented technology is embodied or disembodied. For embodied patents, the rent is generated by sales of the patented product or by a drop of production costs resulting from the new production process. Thus, the decrease of the rent directly reflects the intuitive idea that competition with newer inventions progressively erodes these sources of benefits as the patented product or process ages. The rent that accrues from a disembodied patent is rather linked to indirect pecuniary advantages. It may result, for instance, from a signalling effect that makes the external financiers of a patenting firm more confident about its ability to innovate and generate cash flow in the future. Maintaining a patent in its disembodied form for a long period alters the quality of the signal and the subsequent pecuniary advantages, causing a decrease of the associated rent. Note that the switch from the disembodied form to the embodied one is based on the relative values of the rent in the two respective forms and is thus not inconsistent with the fact that the rent decreases with the age of the patent whatever the form considered. The decrease of the rent and the increase of renewal fees imply that the exercise rule

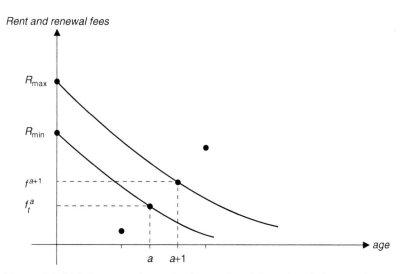

Figure 5.1 Link between the optimal age of withdrawal and the initial rent under the assumption of a single crossing-point of the profile of the rent and the profile of renewal fees.

in the option problem (5.7) can be dramatically simplified as follows: 'Renew the patent as long as the rent exceeds the renewal fee and withdraw the patent otherwise.' The main interest of this simplified exercise rule is that, once coupled with the assumption of a constant and identical rate of decrease for the rent of all patents, an interval of values can then be directly associated to the observed age of with-drawal. This is illustrated by Figure 5.1 where the combinations of age and renewal fees are represented by the different points. Consider a patent that has been renewed up to age a but is withdrawn at age $a + 1$. It means that the rent at age a was greater than the renewal fee at this age and conversely at age $a + 1$. We are able to determine the minimal rent $R_{\min}$ and the maximal rent $R_{\max}$ that are consistent with respectively a renewal at age a and a withdrawal at age $a + 1$. Indeed, given that the rate of depreciation is fixed for all patents, we deduce the two bounds of the initial rent from the renewal fees f_t^a and f_t^{a+1}. These bounds are the ordinates at the origin associated with the two time-paths of the rent represented by continuous decreasing curves in Figure 5.1.

Note that monetary bounds for the initial rent of patents are obtained in Figure 5.1 because we know the value of renewal fees. If renewal fees were not observed, we could assume that a patent is withdrawn if and only if the rent net of all costs of maintaining the patent alive is positive and we would be able to rank initial rents on this basis. However any change of units in the value of bounds separating the different intervals would be consistent with observed ages of withdrawal. Consequently, it would be possible to rank patents according to their initial value but not to assess their monetary value.

5.3.3 Results and perspectives

Empirical studies on patent renewal decisions were initially intended to get some insights into the overall distribution of a population of patents. As already mentioned, a first strand of studies based on the general option model defined in (5.7) was initiated by Pakes (1986). These studies essentially provide empirical evidence that supports the assumption of a decreasing rent. Indeed Pakes (1986) and later on Lanjouw (1998) with a similar model find that most learning about the market is over by the fifth age of protection and that few patents yield higher returns after that point. These studies also confirm that the distribution of patent values is highly skewed. A second strand of studies based on the assumption of a decreasing rent was developed by Schankerman and Pakes (1986) and continued by Schankerman (1998), Deng (2007) and Grönqvist (2009) with the same goal of characterising the distribution of patent values. The econometric model initially proposed by Schankerman and Pakes (1986) is adapted to aggregated data reporting observed frequencies of withdrawal at the different ages of patents for a cohort. The model also works on subsets of patents provided that no source of heterogeneity across these patents is considered. The distribution of patents values inside a single technological field has typically been examined thanks to this model. Nevertheless, as Schankerman and Pakes (1986) note, the next step in econometric works dealing with patent values is 'to examine the empirical characteristics and the theoretical determinants of variation in the quality dimension at a more disaggregated level among different industries and between different types of patents'. This further step has been taken by Barney (2002) and more recently Bessen (2008) and Baudry and Dumont (2012) who all introduce patent metrics as a source of heterogeneity. Both Barney (2002) and Bessen (2008) propose a discrete choice ordered Probit model that is directly inspired by Figure 5.1 and very close in its form to the model presented in (5.2) for stated value approaches. They assume that the initial rent is drawn from a Gaussian distribution with an expected value expressed as a functional form of patent metrics and a standard deviation that takes a similar value for all patents. The constant and identical depreciation rate that affects the rent of patents is estimated simultaneously with coefficients associated to patent metrics by Bessen (2008). The bounds of the interval of values for the initial rent are functions of all these parameters. For his part, Barney (2002) poses the simplifying assumption that the rent is constant over ages and directly infers the bounds of the interval of values for the initial rent from the level of renewal fees. By contrast with Barney (2002) and Bessen (2008), Baudry and Dumont (2012) not only consider heterogeneity in the initial rent but also in the depreciation rate and develop a discrete time duration model for this purpose. They argue that allowing for heterogeneity in the dynamics of the rent is the only consistent way to capture the impact of forward citations. Indeed, forward citations counts evolve as a patent ages and cannot be introduced as static variables like claims or backward citations. Moreover, they show that coupling heterogeneity of the initial rent and of the dynamics of the rent may yield a different ranking of patents depending on whether they are compared in terms of their duration or in terms of their economic value. They also allow for some stochastic component in the dynamics of the rent and find empirical evidence of its importance. The patent metrics used by Barney (2002) include the number of claims, the average number of words per independent claim, the length of written specification, the number of recorded priority and the number of forward citations at the fourth year of life. Unfortunately, estimated coefficients are not reported. Bessen (2008) restricts his analysis to the number of claims, backward citations and forward citations but controls for multiple characteristics of the patent owner. In addition to estimated coefficients, Bessen (2008) also evaluates the portion of total variance in patent initial rents that can be explained by citations statistics. He finds that '*in no case did the portion of variance explained equal as much as 6%. In other words as other researchers have also concluded, patent citation statistics are correlated with patent value, but they are very noisy signals.*' For their part, Baudry and Dumont (2012) consider the following metrics as sources of heterogeneity in the initial rent: the number of claims, the number of keywords, the number of priority, the number of four-digit subclasses of the IPC declared in the patent application, two measures of the degree of specialisation of patents also based on declared IPC codes, the size of family, the number of backward citations made by the applicant on the one hand and by the examiner on the other hand and, finally, the number of countries designated by the patent.[12] They also control for some characteristics of patent holders. The only metric used as a source of heterogeneity in the dynamics of the rent is the number of forward citations. Note however that counts of forward citations are updated for each age of patents. They conclude that patent metrics significantly contribute to discriminate among patents but not enough to obtain a reliable scoring because unobserved heterogeneity predominates.

5.4 Synthesis

Econometric methods that use patent metrics to analyse and explain the heterogeneity of patents as regards their economic value rely on different strategies. None of the methods described in this chapter is perfect. Stated value approaches are costly to implement and to update and may potentially be subject to biases. However, as shown in Table 5.1, several metrics have a highly significant impact on stated values. In essence, no additional information compared to that available to financiers is produced by studies based on indirect market value, more specifically Tobin's Q studies. Moreover, Table 5.1 stresses that only few patent metrics are dealt with by these studies and that their significance level may greatly vary from one study to another.[13] The resulting patent scoring cannot therefore be thought of as an efficient or reliable tool that can help to better discriminate among patents. Like revealed value approaches they do not currently distinguish embodied and disembodied patents. Revealed value approaches for their part do not account for the strategic component of the value of patents. Results on the significance of patent metrics are ambiguous. Bessen (2008) obtains high significance

---

[12] This metric is specific to patents applied for at the EPO. Applicants to the EPO have to declare which member countries are targeted by their patent and have to pay national renewal fees in each of these countries once the patent is granted by the EPO. Lanjouw, Pakes and Putnam (1998) already used the number of applications in different countries as an indicator of quality to weight patent counts.

[13] Significance levels for the study by Lanjouw and Schankerman (2004) are based on *t*-statistics obtained in Tobin's Q equation for the weighted count of patents with the values of quality index as weights.

levels but considers only a limited number of metrics. By contrast, Baudry and Dumont (2012) consider a large set of metrics but obtain limited significant effects. Barney (2002) for his part does not report *t*-statistics. In spite of these differences, a striking feature of these different methods is that they all point to the same conclusion as regards the feasibility of automated patent scoring. Indeed, they all conclude that the role of patent metrics in explaining the total variance of patent values is rather limited compared to that of unobserved sources of heterogeneity.

References

Austin, D. H., 1993. An event-study approach to measuring innovative output: the case of biotechnology. *American Economic Review*, 83, 253–258.

Barney, J. A., 2002. A study of patent mortality rates: using statistical survival analysis to rate and value patent assets. *AIPLA Quarterly Journal*, 30, 317–352.

Baudry, M. and Dumont, B., 2006. Patent renewals as option: improving the mechanism for weeding out lousy patents. *Review of Industrial Organization*, 28, 41–62.

2012. *Valuing Patents Using Renewal Data: An Inquiry into the Feasibility of an Automated Patent Scoring Method*, EconomiX Working Papers 2012–31. http://economix.fr/pdf/dt/2012/WP_EcoX_2012-31.pdf.

Bessen, J., 2008. The value of U.S. patents by owner and patent characteristics. *Research Policy*, 37, 932–945.

2009. Estimates of patent rents from firm market value. *Research Policy*, 38, 1604–1616.

Bloom, N. and Van Reenen, J., 2002. Patents, real options and firm performance. *Economic Journal*, 112, 97–116.

Cockburn, I. and Griliches, Z., 1988. Industry effects and appropriability measures in the stock market's valuation of R&D and patents. *American Economic Review*, 78, 419–423.

Deng, Y., 2007. Private value of European patents. *European Economic Review*, 51, 1785–1812.

2011. A dynamic stochastic analysis of international patent application and renewal processes. *International Journal of Industrial Organization*, 29, 766–777.

Gambardella, A., Harhoff, D. and Verspagen, B., 2008. The value of European patents. *European Management Review*, 5, 69–84.

2013. *The Economic Value of Patent Portfolios*, Discussion Paper 9264. London: Centre for Economic Policy Research.

Goddar, H. and Moser, U., 2011. Traditional valuation methods: cost, market and income approach, in *The Economic Valuation of Patents*, eds. Munari, F. and Oriani, R., pp. 109–140. Cheltenham, UK: Edward Elgar.

Greene, W. H., 1993. *Econometric Analysis*, 3rd edn. New York: Prentice Hall.

Griliches, Z., 1981. Market value, R&D and patents. *Economics Letters*, 7, 183–187.

Grönqvist, C., 2009. The private value of patents by patent characteristics: evidence from Finland. *Journal of Technology Transfer*, 34, 159–168.

Hall, B. H. and MacGarvie, M., 2010. The private value of software patents. *Research Policy*, 39, 994–1009.

Hall, B. H., Jaffe, A. and Trajtenberg, M., 2005. Market value and patent citations. *RAND Journal of Economics*, 36, 16–38.

Harhoff, D., Narin, F., Scherer, M. and Vopel, K., 1999. Citation frequency and the value of patented inventions. *Review of Economics and Statistics*, 81, 511–515.

Harhoff, D., Scherer, M. and Vopel, K., 2003. Citations, family size, opposition and the value of patent rights. *Research Policy*, 32, 1343–1363.

Hayashi, F., 1982. Tobin's marginal q and average q: a neoclassical interpretation. *Econometrica*, 50, 213–224.

Hayashi, F. and Inoue, T., 1991. The relation between firm growth and q with multiple capital goods: theory and evidence from panel data on Japanese firms. *Econometrica*, 59, 731–753.

Lanjouw, J., 1998. Patent protection in the shadow of infringement: simulation estimations of patent value. *Review of Economic Studies*, 65, 671–710.

Lanjouw, J. and Schankerman, M., 2004. Patent quality and research productivity: measuring innovation with multiple indicators. *Economic Journal*, 114, 441–465.

Lanjouw, J., Pakes, A. and Putnam, J., 1998. How to count patents and value intellectual property: the uses of patent renewal and application data. *Journal of Industrial Economics*, 46, 405–432.

Morricone, S. and Oriani, R., 2011. Stock market valuation of patents portfolios, in *The Economic Valuation of Patents*, eds. Munari, F. and Oriani, R., pp. 337–363. Cheltenham, UK: Edward Elgar.

Omland, N., 2011. Valuing patents through indicators, in *The Economic Valuation of Patents*, eds. Munari, F. and Oriani, R., pp. 169–201. Cheltenham, UK: Edward Elgar.

Pakes, A., 1986. Patents as options: some estimates of the value of holding European patent stock. *Econometrica*, 54, 755–784.

Schankerman, M., 1998. How valuable is patent protection? Estimates by technology field. *RAND Journal of Economics*, 29, 77–107.

Schankerman, M. and Pakes, A., 1986. Estimates of the value of patent rights in European countries during the post-1950 period. *Economic Journal*, **96**, 1052–1076.

Silverberg, G. and Verspagen, B., 2007. The size distribution of innovations revisited: an application of extreme value statistics to citation and value measures of patent significance. *Journal of Econometrics*, **139**, 318–339.

Suzuki, J., 2011. Structural modelling of the value of patent. *Research Policy*, **40**, 986–1000.

6 | Dysfunctions of the patent system and their effects on competition

DAVID ENCAOUA AND THIERRY MADIÈS

6.1 Introduction

The objective of this chapter is to analyse some current conflicts between patents and competition. Analysis of the tensions between these two instruments has been the subject of a large body of the economic literature and, after a long academic debate, patents and market competition are now considered by most economists as offering complementary rather than substitutable incentives in favour of innovation.[1] Two statements summarize the current economic beliefs concerning the relationships between patents and competition: on the one hand, firms innovate to escape from competition (Darwinian view); on the other hand, the social benefits of innovation cannot be captured without the spur of competition and the scrutiny of antitrust law (mainstream view). However, despite this favourable convergence, some important tensions persist between the intellectual property and the antitrust authorities. One of the main arguments developed in this chapter is that these tensions mainly occur because the current patent system presents some dysfunctions that create distortions difficult to solve under the antitrust rulings. Besides the traditional conflict between dynamic efficiency and static deadweight loss, the contemporary dysfunctions of the patent system

We thank Simon Lapointe for a careful reading of a previous version and useful suggestions.

[1] The question whether patents and competition are complementary or substitutable instruments of the innovation process has been debated for a long time. Since the first Schumpeterian analyses and the more recent works of Bessen and Maskin (2009), Boldrin and Levine (2008), various economists (see Aghion *et al.*, 2001; Aghion and Griffith, 2005; Encaoua and Ulph, 2005) have shown that in a dynamic rivalry framework with a step-by-step process of innovation, the two instruments tend to be complementary as long as the intensity of competition in the product market is not too high. Moreover, the contemporary evolution of competition policy towards intellectual property is largely favourable to the complementary perspective.

can be made easier to understand by recalling how the economic representation of patents has evolved.

In its initial and most common representation, a patent protects a discrete innovation, in other words an isolated innovation specific to a technology, without links to any downward or upward technology. It confers on its holder the temporary power to stop a third party from using the protected information that is disclosed in the patent without the holder's consent. A large proportion of the initial literature on patents uses the discrete innovation framework and reaches the unsurprising conclusion that a stronger patent protection promotes innovation. To obtain this result, the literature implicitly makes a number of assumptions.[2] In particular, it assumes that a patent is an intellectual property asset satisfying the following properties: (1) it confers a perfect protection to its holder: once granted it is presumed to be an unquestionable right. Moreover, it perfectly defines the protected claims that are disclosed in the patent. In other words, even though it is an intellectual property right, a patent is treated as any other iron-clad property right; (2) a patent is supposed to be the best instrument of protection because, unlike trade secrecy, it allows the holder to collect damages when an infringement is detected; and (3) a patent delivers an autonomous and independent piece of information about knowledge, in the sense that the innovation it protects is not linked to other innovations, either upstream or downstream. In other words, there is a clear and unique connection between a technology and a patent. All these implicit assumptions were instrumental in the development of a huge economic literature, focused on one hand on the socially optimal design of the patent in terms of duration and scope of protection, and on the other hand on the links between innovation and competition (Gilbert, 2006). The progressive challenge to these different assumptions has contributed to the analysis of new concerns raised by the development of the patent system.

The challenge to the first assumption has led to a careful (re)examination of the consequences of a rather uncertain protection by the patent. The uncertainty has been exacerbated in recent years by abusive award of patents to applications that do not fully satisfy the patentability criteria, notably that of the required inventiveness (or non-obviousness in American usage). Awarding patent rights to

---

[2] For an overview, see Encaoua et al. (2006) and Harhoff et al. (2007).

dubious applications is thus the first dysfunction of the patent system. It considerably affects competition in the product market. Patents that are dubious or too broad (i.e. incorporating claims with an excessive scope) harm the equilibrium between intellectual property and competition by discouraging the monitoring of innovation and by artificially increasing final prices due to unwarranted royalties. In its 2003 report, the Federal Trade Commission concluded that increasing the quality of awarded patents should be a top priority to re-establish the equilibrium between intellectual property and competition (FTC, 2003). When uncertain patents covering valuable products exist, it is no longer true that stronger protection promotes innovation. Moreover, uncertain protection undermines the intrinsic superiority of the patent over other forms of protection, suggesting that if the number of patent applications is maintained at a very high level, it is because a patent has a different function than simply ensuring protection against imitation.

A second characteristic of the patent system is the fact that patents are increasingly used as instruments in technology exchanges. It is illustrated by the importance of the open innovation organization and the frequent use of patent trading between various agents.[3] Assumption (3) in the traditional representation of the patent is now replaced by a cumulative conception of the innovation process. The view that technological change is not an isolated event but a sequential process in which each innovation is built on knowledge patented by forerunners has two consequences. First, the patent's function is not only to protect the holder from a malicious imitator (backward protection), but also to block a follow-up innovation (forward protection). Second, while the discrete view emphasized competition in the product markets, the cumulative view focuses on competition on the markets of technological exchanges. Technological competition in these markets may be affected by 'hold-up' behaviour. For example, a patent holder can accuse a producer of patent infringement, by pretending that the patent covers some of the technologies that he is using for his production. This accusation can be especially detrimental to the producer when it is made *ex post*, i.e. after the second innovator

---

[3] The contemporary specificities of the modes of organization and coordination of research put the emphasis on open innovation systems, in which knowledge exchange between multiple agents takes priority over independent and individual research.

has completed the investment necessary for his activity. A further escalation is reached when the patent holder sues for infringement not on a delivered patent, but on a patent pending not yet published. Indeed, the laws on divisional patent applications or on continuation patent applications authorize an initial depositor to act as if the date of the new application ran from the date of the initial application. In some situations, the potential infringer could have acted while ignoring the existence of a divisional patent application, which would explain why he did not request an authorized license from the holder before starting production of his own good. Finally, when a license royalty is required *ex post* by the patent holder, its level can be excessive compared to the level that would have been required if the negotiation had happened *ex ante* (i.e. before the producer started production of his good). The hold-up situation is made worse when the patent holder requests that the potential infringer stops his activity by obtaining a relief injunction. The threat of injunction that weighs on the downstream innovator clearly gives an advantage to the upstream innovator. Are such threats legal from the point of view of intellectual property law? Some believe so, arguing that an injunction is the normal consequence of the protection awarded to a patent holder. Others believe that injunctions should not be authorized, arguing that these practices constitute serious obstacles to the continuous operation of markets.[4]

Finally, a third and more contemporary dimension of the patent system further complicates the question of the tensions between patents and competition. In the cumulative conception of innovation, some patents act not only in a sequential manner as assumed in the previous paragraph, but also in clusters: it is the case when the use of numerous independent patents is necessary to produce a single good.[5] Not only does this clustering increase the transaction costs due to the

---

[4] A question that appears essential is whether competition authorities have the appropriate legal instruments to convict abusing patent holders. For example, it is possible that some behaviour can be considered abusive while its author is not in a dominant position, which makes the application of antitrust laws difficult at least in Europe.

[5] Scotchmer (2004) uses the term *research tools* to designate the set of patented inputs that are involved in the production of a second-generation good. For example, the cultivation of genetically modified seeds simultaneously requires the use of different genes, each with its specific properties. It also requires potentially patented techniques to insert these genes in the germplasm (the element issued from the vegetal source that can give a new individual).

so-called 'patent thicket', but it may also create an obstacle to further innovation. It is now the fight *for* the market, and not *on* the market, that matters. Patent races are essential instruments in this competition. In opposition to the free-entry assumptions used in most of the endogenous growth models, patent races are governed by imperfect competition. The accumulated patents strongly influence the structure of the races, thus having an effect on the competitive mechanisms on the innovation markets. The massive sales of patent portfolios that are observed illustrate the important role of these intellectual property instruments for the conquest of new markets. Two factors contribute to the appearance of this phenomenon: the conjunction of a high fragmentation of intellectual property rights, and the development of a great number of complex technologies that require many patented elementary components. Many sectors of the information and communication technology (ICT) industries produce goods that require the use of a great number of patented components. Even if each component represents only a minor part of the value of the final good, it still may be an essential input. In part because the coordinated price of a collective licence can be inferior to the sum of the prices of the independent licences, economic efficiency requires that complementary patents be coordinated through a patent pool or through a Standard Setting Organization (SSO) tasked with the choice of a technological standard. However, as in any process in which complementarities between independent inputs require a coordinated behaviour, individual deviations from the collectively defined norms are frequent. For example, it may happen that during an SSO's deliberation, an agent certifies that he has no intellectual property relevant to the standard. Thus, the chosen standard may be based on this certification. But after the standard adoption, the same agent allegedly demands royalties from those using its technology in connection with that standard. This type of deviation, illustrating once again hold-up behaviour, can have significant impacts in terms of competition, especially in the innovation market. The coordination process can also cause other different problems. For example, the determination of the essential patents can lead to conflicts when substitutable technologies are in presence. Moreover, agreeing on the economic principles that a collective licence price must satisfy is not an easy task. Finally, the risk of collusive behaviour is never absent. Indeed, it is difficult for antitrust authorities to mitigate this risk.

This chapter is organized around the three abovementioned dysfunctional aspects that affect the competition process respectively in the product markets, the technology markets and the innovation markets: the issue of bad-quality patents (Section 2), hold-up behaviour in follow-up innovation (Section 3), and the coordination issue in patent pools and technological standards (Section 4).

6.2 Patent quality and competition in the product markets

It is commonly recognized that the number of patent applications is very large and increasing, as illustrated by the following figure for the United States: 'Every Tuesday, the day of the week the US Patent and Trademark Office (USPTO) issues new patents, there are roughly 3,500 new patents, corresponding to new IP rights that no American is allowed to infringe, and for which there is no fair use defence to patent infringement like with copyright and trademark' (Lei and Wright, 2010). This figure reflects the sign of a very dynamic and innovative economy as that of the United States but one cannot disregard the possibility that many of these patents may be of bad quality, in the sense that they fail to satisfy at least one of the patentability criteria: utility, novelty and non-obviousness (inventivity). The US Patent and Trademark Office (USPTO), and to a lesser extent the European Patent Office (EPO), are trapped in an uncomfortable position defined as follows by Lemley (2012):

The Patent and Trademark Office finds itself caught in a vise. On the one hand, it has been issuing a large number of dubious patents over the past twenty years, particularly in the software and electronic commerce space. It issues many more patents than its counterparts in Europe and Japan (Van Pottelsberghe de la Potterie, 2010); roughly three-fourths of applicants ultimately get one or more patents, a higher percentage than in other countries. Complaints about those bad patents are legion, and indeed when they make it to litigation they are quite often held invalid. Even the ones that turn out to be valid are often impossible to understand; in the information technology industries, there is no lawsuit filed in which the parties don't fight over the meaning of patent claim terms (Bessen and Meurer, 2008).

It is now largely admitted that a patent is not an iron-clad right as are other forms of property. A patent is more likely an uncertain or a probabilistic right (Ayres and Klemperer, 1999; Lemley and Shapiro, 2005; Shapiro, 2003). The two most important points are that some of the dubious patents are of large economic significance and the enforcement uncertainty is largely strengthened by the issuance of too many bad quality patents.

Three questions emerge: (1) Why are so many bad-quality patents granted by a patent office? (2) How does uncertainty about the patents' quality affect the choice of a protection regime? (3) To what extent do the patent holder and potential infringers prefer to settle their private disputes on the patent's validity rather than pursue their litigation in court, and what are the consequences of these private settlements on market competition?

6.2.1 Two views of the reasons for the issuance of bad-quality patents

Besides the anecdotal evidence on some exotic patents,[6] we know that a large number of economically valuable products are protected by weak patents, i.e. patents that would probably be invalidated by a court if they were the subjects of litigation. Examples include drugs such as Lipitor, genetic material such as the stem cells of the Wisconsin Alumni Research Foundation (WARF), the breast cancer genes *BRCA1* and *BRCA2* (Myriad Genetics), software such as the File Allocation Table (FAT) by Microsoft, etc.[7]

Why are so many bad-quality patents granted by the USPTO? To answer this question, Lei and Wright (2010) contrast two possible explanations. The first relies on the 'rational ignorance' principle advanced by Lemley (2001), according to which the USPTO examiners voluntarily conduct insufficient prior art search that could render weak patents unpatentable. Lemley argues that by so doing, US examiners 'are "rationally ignorant" of the objective validity of patents ... because it is too costly for them to discover those facts'. Given the skewed nature of the patent value distribution, 'society would be better off economizing on examinations, deferring rigorous determination of validity until the

---

[6] Examples include crustless peanut butter, jelly sandwich, swinging on the swing, etc. See Jaffe and Lerner (2004).

[7] See the website www.pubpat.org of the Public Patent Foundation (PUBPAT). This foundation represents the interests of the public in the patent system. Even if the foundation shares the view that a properly functioning patent system can help an innovative economy, it takes great care to avoid the negative effects that over-patenting, unmerited patenting and excessive patent rights can have on society. Its aim is to protect freedom from illegitimate constraints.

patent enters litigation'. The second explanation denies the examiner's ignorance and focuses on the existence of an 'institutional pro-applicant bias' in the USPTO. 'There exists a pro-applicant bias of policies and procedures at the USPTO that renders patent examiners' effort useless: they are encouraged by various institutional incentives to accept applications that they nevertheless perceive to be ineligible' (Jaffe and Lerner, 2004). There are at least three institutional biases that force the USPTO examiners to grant undeserved patents: (i) burden of proof, (ii) incentives and (iii) continuation and divisional patents.

(i) Burden of proof

In the United States as in Europe, an applicant does not have to prove that the application is patentable. It is the examiner's burden to prove that the application is unpatentable, which is generally a very time-consuming task. The opposition procedure in Europe allows a third party to make an opposition. This procedure is adversarial between the challenger and the patent holder. By contrast, in the United States, the re-examination procedure does not involve any adversarial procedure since it maintains an exclusive relationship between the applicant and the patent examiner (Graham *et al.*, 2004; Harhoff and Reitzig, 2004).

(ii) Incentives

Patent examiners in the USA are mainly rewarded on patents granted, and do not bear the aftermath of granting questionable patents. 'The salaries of US examiners are tied to the number of applications they process: they have production quotas to meet, and earn bonuses when they exceed their quotas by at least 10% ... Importantly, they are never liable in the event patents are invalidated in court and there are no negative consequences for examiners who produce low-quality work' (Langinier and Marcoule, 2009). In Europe, the rewards of EPO examiners are equivalent to those at the USPTO, but the tie between salary and productivity is somewhat less close. This difference is illustrated by a lower workload for European examiners than for American ones: the number of filings per examiner is 37 at the EPO and 97 at the USPTO, and the pendency allowed for both the prior art search and the examination process is 50 months at the EPO versus 27 months at the USPTO. These figures may explain why the EPO grants fewer dubious patents than the USPTO, and illustrate the importance of the quantity–quality trade-off in the patent system (Guellec and van Pottelsberghe de la Potterie, 2007, ch. 7).

(iii) Continuation and divisional patents

Continuation and divisional patents are common practices.

A continuation is a second application for the same invention claimed in a prior nonprovisional application and filed before the original prior application becomes abandoned or patented. A divisional application or division is a later application for an independent or distinct invention, carved out of a pending application and claiming only subject matter disclosed in the earlier or parent application. Continuations are not permitted by the EPO and the JPO but are allowed by the USPTO. Divisional patents are permitted by the EPO, and provide a kind of alternative to continuations (Hedge *et al.*, 2007).

Such applications offer means to tune a patent's claims to changing circumstances. According to Lemley (2012), in the USA, the use of continuations is largely applicant- rather than examiner-driven. Moreover, the ability of applicants to file an unlimited number of continuation applications – and their willingness to do so – makes it difficult for examiners to simply reject bad applications. 'Applicants view a rejection as simply a negotiating position that invites a counteroffer, not as a judgment that their application is in fact unpatentable. And because they can continue making counteroffers, increasing the number of rejections simply prolongs the application process' (Lemley, 2012).

In order to test whether the rational ignorance principle or institutional bias best explains the issuance of bad-quality patents in the USA, Lei and Wright (2010) construct a sample of twin patents, i.e. patents that were filed to both the USPTO and the EPO during the period 1990–95 and were granted by the USPTO while resulting at the EPO in one of the three possible outcomes: granted, rejected or withdrawn by the applicant. They assume that the EPO grants fewer dubious patents than the USPTO and they test whether the rate of failure at the EPO is linked to the prior art research effort made by US examiners. The outcome of a patent prosecution at the EPO is used as an indicator of the patents' strength: an application for a weak patent would have a high probability to be either withdrawn by the applicant or rejected by the EPO. In order to measure the prior art research effort made by USPTO examiners, the authors define the *prior patents search intensity* (PPSI) by the ratio $PPSI = CPP/(CPP + UPP)$, where the variable CPP is the number of *cited prior patents*, and the variable UPP is the number of *uncited prior patents*, both evaluated for each patent. While the number of *cited prior patents* (CPP) appears directly in the application, the number of *uncited prior patents* (UPP) was computed by the authors by using a specific algorithm (Latent Semantic Analysis). The higher the value of the PPSI ratio, the higher is the research effort devoted by the examiners to find the appropriate prior art. The purpose is to test which one of the two following alternative hypotheses H_1 and H_2 best explains the issuance of weak patents by the USPTO, i.e. patents that are withdrawn or rejected by examiners at the EPO:

H_1: *Rational ignorance:* a USPTO patent with a high amount of cited prior art (i.e. a patent having a high PPSI) signals a *strong* patent that would have a high probability of being accepted by the EPO.

H_2: *Institutional bias:* a USPTO patent with a high amount of cited prior art (i.e. a patent having a high PPSI) signals a *weak* patent that would have a high probability of being rejected by the EPO or withdrawn by the applicant while it is granted by the USPTO.

The main econometric result of Lei and Wright is in favour of H_2: the probability of failure at the EPO is significantly and positively affected by the PPSI ratio, measuring the research intensity of the prior art at the USPTO. This result suggests that US examiners devote important prior research intensity to patents that they perceive as being weak, but despite their negative perception, the rules and procedures of the USPTO force them to grant a patent to these weak applications. In other words, even though US examiners ultimately fail to reject weak patents, their revealed evaluation measured by the PPSI ratio is a significant predictor of the application outcome at the EPO.

Lemley (2012) suggests two ways to improve, at least partially, the quality of granted patents: a tiered review process and a post-grant opposition procedure:

To harness information in the hands of patent applicants, we could give applicants the option of earning a presumption of validity by paying for a thorough examination of their inventions. Put differently, applicants should be allowed to 'gold plate' their patents by paying for the kind of searching review that would merit a strong presumption of validity. An applicant who chooses not to pay could still get a patent. That patent, however, would be subject to serious – maybe even de novo – review in the event of litigation. . . Post-grant opposition is a process by which parties other than the applicant have the opportunity to request and fund a thorough examination of a recently issued patent. A patent that survives collateral attack should earn a presumption of validity similar to the one available through tiered review. The core difference is that the post-grant opposition is triggered by competitors – presumably competitors looking to invalidate a patent that threatens their industry. Like tiered review, post-grant opposition is attractive because it harnesses private information; this time, information in the hands of competitors. It thus helps the PTO to identify patents that warrant serious review, and it also makes that review less expensive by creating a mechanism by which competitors can share critical information directly with the PTO. A post-grant opposition system is part of the new America Invents Act, but it won't begin to apply for another several years, and the new system will be unavailable to many competitors because of the short time limits for filing an opposition.[8]

6.2.2 Choice of protection regime: 'big secrets' and 'little patents'

A first consequence of the uncertainty attached to a patent is related to the choice made by an inventor to protect his innovation. Empirical studies from inventors' surveys (Arundel and Kabla, 1998; Cohen *et al.*, 2000; Levin *et al.*, 1987) revealed that in many industrial sectors, resort to patents was not the preferred form of protection of innovators. What drives the individual choice between patent or other forms of protection?[9] Some theoretical studies tried to answer this question by taking into account the strength of a patent, evaluated as the probability that a court would confirm both the patent's validity and the infringement. For a strong patent, the probability is close to one, while for a weak patent, it is close to zero. Consider a process innovation that allows the reduction of production costs. The approach developed by Anton and Yao (2004) involves the asymmetry of information between the inventor and third parties: the inventor knows precisely the amount of cost reduction achieved by the

---

[8] Post-grant opposition is available only for patents granted on applications filed after 16 April 2013. Because the average time to grant is close to four years, it will likely be the latter part of this decade before many patents are eligible for post-grant opposition.

[9] See on this question the recent survey by Hall, Helmers and Sena (2012).

innovation, but third parties, including potential competitors, have only partial information. The information that reaches them depends on what the innovator chooses to disclose when patenting his innovation. The cost reduction level disclosed in the patent thus becomes a strategic choice for the patent applicant.[10] However, the imitator can also obtain additional information by himself if he chooses to reverse engineer the innovation. Anton and Yao (2004) show that choosing the patent protection leads to two opposite effects. On one hand, there is an imitation effect: a weaker patent will contain less information, because the potential imitator has a lower chance of being convicted and fined by a court when the patent is weak. Therefore, since it is in the interest of the imitator to copy everything disclosed in the patent by the holder, a weak patent holder's interest is to disclose as little as possible. On the other hand, there is a signal effect: a strong patent encourages a high level of disclosure. Indeed, because a strong patent is unlikely to be imitated, it is not risky for the holder of a strong patent to disclose the effective cost reduction, thus signalling his technological advance to competitors. Obviously, this reasoning would not hold with a weak patent. Overall, the level of information disclosed allows a compromise between the two effects described above. Using a signalling game framework, Anton and Yao (2004) show that in equilibrium, only those process innovations with a cost reduction below a certain threshold stand to gain from being patented, while innovations with value above this threshold benefit more from being kept secret. The main result of this model, in which patent protection is probabilistic, is that the compromise between patent and secret leads to the patenting issue solely for low-value innovations (the so-called 'little patents') which have a small chance of being imitated. In contrast, high-value innovations are better protected by secrecy ('big secrets'). Note that in the model of Anton and Yao (2004), the decision to imitate or not to do so results from the amount of information the innovator chooses to disclose, and not from the independent imitator's choice.

---

[10] It is important to note that even if under the patent law an applicant is required to disclose sufficient information to enable someone skilled in the art to make and use the claimed embodiments of the innovation, what the applicant actually discloses remains discretionary. It may be only a part of the technically relevant information, since it is only what the patent holder discloses in the patent that enables a competitor to imitate.

An alternative approach is developed in Encaoua and Lefouili (2005). In that model, the decision to imitate or not is assumed to depend directly on the imitator's behaviour and not on the innovator's choice. The emphasis is put on three factors: (1) the strength of the patent as previously defined; (2) the relative cost of imitation, whether the innovation is patented or kept secret, the cost of imitation being lower for a patent due to the compulsory disclosure; and (3) the size of the innovation defined as the cost reduction allowed by the process innovation. Furthermore, the level of imitation is an endogenous variable chosen by the potential imitator. Then, the interactions between these different factors lead to novel situations. For example, the choice of patenting the innovation leads to a level of imitation that can differ from the level that would have arisen if the innovation had been kept secret. Moreover, the innovator can be better off when imitated if the loss he suffers on the market from being imitated is more than compensated by the expected value of damages collected.

Two effects are thus identified.[11] The first, called the damages effect, says that with identical levels of imitation, the innovator will always prefer a patent to trade secret, because the imitation of a patent leads to compensation in the form of damages. The second effect, called the competition effect, makes more explicit the influence of the imitation level on the innovator's profit. As stated above, this level of imitation can differ according to the mode of protection. If the level of imitation is lower for a patented innovation than for a secret innovation, the competition effect reinforces the damages effect: the innovator will prefer a patent. However, if the imitation level is higher for a patented innovation than for a secret innovation, the two effects go in opposite directions. The composition of the two effects necessitates in this case a deeper analysis. Adopting the expression of damages

---

[11] The two effects that determine the choice of the protection method are highlighted in the following decomposition. Let $\Pi_{1P}(d_1, d_2, \theta)$ denote the profits induced by a patented innovation when the innovator reduces costs by d_1, the imitator reduces its costs by d_2 and the strength of the patent is θ. Also let $\Pi_{1S}(d_1, d'_2)$ be the profit induced by an innovation kept secret when the innovator reduces costs by d_1, the imitator reduces its costs by d'_2. The difference in profits for the innovator in the two cases can be written as: $\Pi_{1P}(d_1, d_2, \theta) - \Pi_{1S}(d_1, d'_2) = (\Pi_{1P}(d_1, d_2, \theta) - \Pi_{1S}(d_1, d_2)) + (\Pi_{1S}(d_1, d_2) - \Pi_{1S}(d_1, d'_2))$. The first term of the decomposition, which is always positive, corresponds to the damages effect. The second term corresponds to the competition effect. It can be positive (if $d'_2 > d_2$) or negative (if $d'_2 < d_2$).

that result from the unjust enrichment doctrine, Encaoua and Lefouili (2005) reach the following conclusion: in the perfect equilibrium of the three-stage game, where the choice of protection is made at the first stage, the decision to imitate and the choice of the imitation level are made at the second stage, and Cournot competition on the product market occurs in the third stage, there exists a threshold such that the innovations involving a cost reduction below the threshold are always patented, while those with a cost reduction above the threshold are better protected by secret. The result is consistent with the one obtained by Anton and Yao (2004) albeit in a different framework. It finally challenges the justification for the patent as a universal protection instrument for innovation.[12]

6.2.3 Private settlements

Challenging a patent's validity through litigation is costly and difficult for at least three reasons. First, the patent holder may contractually prevent litigating.[13] Second, the required standard of clear and convincing evidence in the USA to prove invalidity is in general very demanding for the challenger, especially for new patentable subject matters like software.[14] Third, challenging has the dimension of a public good: a firm benefits from a successful challenge initiated by another competitor, since it gets the new technology freely. Therefore, the individual incentives to challenge a patent's validity are low. This is why disputes on weak patents are more frequently solved through private settlements than through a judicial litigation procedure. But not only are patents for which litigation goes to completion in front of a US court invalidated in a high proportion (about 50%, according to Allison and Lemley, 1998), they also include patents of great commercial value.[15] For these three

---

[12] Note that both the models of Anton and Yao (2004) and Encaoua and Lefouili (2005) focus on process innovations. We are not aware of an existing analysis of the choice between patent and secret for product innovations.

[13] For instance, in Japan, Microsoft forced its licensed OEM suppliers to pledge not to file lawsuits on the grounds that Windows infringes a patent right.

[14] This is illustrated by the recent *i4i* v. *Microsoft* case. See Supreme Court of the United States, *Microsoft Corp.* v. *I4I Limited Partnership et al.*, Certiorari to the US Court of Appeal for the Federal Circuit, 9 June 2011.

[15] To illustrate this last point, Chiron's patent on monoclonal antibodies specific to breast cancer antigens was invalidated in 2002 in a suit in which Chiron had sought over $1 billion in damages from Genentech. Another example relates to the drug

reasons, private settlements are often preferred to legal proceedings. But, even if public policy encourages private settlements of legal disputes, it does not follow that all settlements are consistent with the public interest. Gilbert (2006) makes the following distinction:

If the patent is valid and would be infringed, the patent gives its owner the right to exclude a rival that employs the teaching of the patent, and a settlement that allows the alleged infringer to stay in the market would not be anticompetitive. On the other hand, if the patent is not valid or would not be infringed, a settlement between a patentee and a potential entrant that limits the ability of the entrant to compete against the patentee could harm competition that would have occurred in the absence of the settlement.

We discuss briefly two types of private settlements that affect negatively the competition in the product market: the so-called 'reverse payments' practice observed in the pharmaceutical industry and the specific 'per-unit royalty licensing scheme' which occurs under the shadow of patent litigation.

Reverse payments in the pharmaceutical industry

Some branded pharmaceutical companies pay a large amount of money to a generic producer to delay its entry in their market. This practice harms society since it delays access to a less expensive generic drug. However, it is not obvious whether this practice violates antitrust rules.[16]

Prozac. The US Court of Appeals for the Federal Circuit invalidated an Eli Lilly patent on Prozac in 2000 and caused Lilly's share price to drop 31% in a day.

[16] The assessment of this practice by the US Supreme Court is illustrated by this quotation extracted from 'Brief for the US as Amicus Curiae in Federal Trade Commission, *Petitioner v. Schering-Plough Corporation et al.*' (US Supreme Court 05–273): 'Patent litigation settlements that include reverse payments thus implicate conflicting policy considerations and complex legal issues at the intersection of patent and antitrust law, with further complexity introduced in the pharmaceutical context by the dynamics of the Hatch–Waxman Act. On the one hand, the interests in consumer welfare protected by the antitrust laws militate against adoption of a legal standard that would facilitate patent holders' efforts to preserve weak patents by dividing their monopoly profits with settling challengers. The risks are magnified when the settling parties are in a position to utilize the Hatch–Waxman exclusivity period to further constrain competition from other generic manufacturers. On the other hand, the public policy favoring settlements, and the statutory right of patentees to exclude competition within the scope of their patents, would potentially be frustrated by a rule of law that subjected patent settlements involving reverse payments to automatic or near-automatic invalidation.'

On the two sides of the Atlantic, the antitrust enforcers believe that reverse payments are anticompetitive since they improperly raise consumer costs by keeping out less expensive generic drugs. But some differences appear between the USA and the European Union (EU). In the USA, the Federal Trade Commission (FTC) and the courts have radically opposed views on this practice. While the FTC considers such settlements as unlawful regardless of who ultimately would have won the patent litigation, the US courts reject this reasoning, requiring those challenging such reverse payments to show that the settlement impacts competition for products not covered by the patents, or that the underlying patent infringement is objectively baseless or based on fraud. In Europe, both the Directorate-General for Competition and the Court of First Instance (CFI) seem to agree in condemning private settlements that involve a reverse payment, whether the patent is valid or not (as illustrated by the recent *Boehringer Ingelheim* v. *Almirall*). Moreover, commentators differ markedly in their views of reverse payment settlements. Some of them believe that, in the USA, reverse payments are a consequence of the facilitation of generic entry procedure derived from the Hatch–Waxman Act.[17] Others consider that the reverse payment cases constitute an 'important category of cases in which the terms of the settlement themselves tend to indicate that the patent was weak and thus that competition was diminished by the settlement' (Shapiro, 2003).

Licensing a weak patent under the shadow of patent litigation

Consider the situation where the holder of a patented process innovation is confronted to a set of oligopolistic firms that are potential users of the cost-reducing innovation. Suppose that the patent covering this

---

[17] One of the peculiarities introduced by the Hatch–Waxman Act (1984) is to allow a generic drug to avoid the usual tests (known as the NDA, New Drug Application) required by the Federal Drug Agency to obtain the market's entry approval. For a generic, the Hatch–Waxman Act requires a less demanding test (known as the ANDA, Abbreviated New Drug Application). An ANDA requires only demonstrating that the proposed generic drug is bioequivalent to an approved pioneer drug. Providing evidence of safety and effectiveness from clinical data or from the scientific literature is not necessary. To compensate such easier entry by generics, the Hatch–Waxman Act extended the patent length of the brand name drug by restoring the time period lost while awaiting NDA approval. The maximum extension period is capped at a five-year period, or a total effective patent term after the extension of not more than 14 years (see Appelt, 2010; Thomas, 2006).

innovation is weak. Licensing it under the shadow of patent litigation implies that the patent holder prefers to license its patent at a price that deters any challenge rather than the option to pursue an infringer in a court, since in a trial the infringer will probably try to challenge the patent's validity. Therefore the licensing price is such that any potential licensee will prefer to accept the license rather than to litigate the patent's validity. In other words, licensing a weak patent under the shadow of patent litigation implies a licensing scheme that deters any litigation.[18] A recent paper by Amir *et al.* (2011) investigates the nature of the best licensing scheme when the patent is weak and asks whether the result is robust against features that matter when the patent is certain. The main result in Amir *et al.* is that a weak patent holder prefers a per-unit royalty to a fixed fee when the positive price effect outweighs the negative quantity effect of a cost increase. The authors show that this condition is satisfied regardless of whether: (1) the licensor is an outsider or an insider in the downward oligopoly market, (2) the licensees compete à la Cournot or à la Bertrand, and (3) the downward market includes homogeneous or differentiated products. Thus, contrary to what happens for an iron-clad right, the owner of a weak patent always prefers a per-unit royalty to a fixed fee.[19] Licensing a weak patent with per-unit royalty allows the patent holder to extract more revenues than with a fixed fee, because the licensee can pass the corresponding unit cost increase to the consumers, explaining why licensees also prefer a per-unit royalty. The only agents that suffer are the consumers of the final product: they pay a higher price than they would if the patent's assessment was made prior to licensing. Therefore licensing a weak patent under the shadow of

---

[18] A recent OECD inquiry (Zuniga and Guellec, 2009) finds that avoiding patent litigation is an important motive of licensing.

[19] The intuition of this result may be summarized as follows. For a certainly valid patent, the patent holder cannot control the number of licensees when a per-unit royalty is used while he can with a fixed-fee license. This advantage of fixed-fee over per-unit royalty contracts is absent when the holder of a weak patent wants to make a license offer that deters litigation: indeed such offer has to be accepted by *all* firms. This suggests that a per-unit royalty scheme can be preferred over a fixed-fee scheme if the patent is weak whereas the reverse holds if the patent is valid with certainty. Moreover, since licensing occurs *ex ante*, i.e. before an assessment of the patent's validity, the per-unit royalty licensing revenue is greater than the expected revenue obtained when licensing occurs after the patent's validity assessment, whereas fixed-fee licensing leads to the same revenue whether licensing is made *ex ante* or *ex post*.

patent litigation raises both a private problem and a public problem. The private problem arises because individual potential licensees have insufficient incentives to challenge the patent's validity alone: this is a direct consequence of the public good nature of challenging.[20] A public policy problem also arises because a weak patent holder is able to harm consumers by setting a per-unit royalty that increases the cost and therefore the price.

6.3 Sequential innovations, hold-up and competition in technological markets

We have so far examined the case of a discrete innovation and have analysed the effects of the uncertainty relative to the patent's validity by showing how this uncertainty affects the competition on the product market. We now examine the effects of patents covering follow-up or sequential innovations in the sense that each innovation is built on previous patented knowledge. In such a framework, the delimitation of property rights between downstream and upstream innovators becomes crucial (Scotchmer, 2004). The uncertainty inherent to this delimitation is likely to result in a hold-up. For instance, a patent holder can unpredictably block the activities of an existing producer, accusing him of using the technology protected by the patent without the owner's consent.[21] The plaintiff can be either the patent owner himself or some other party acting on behalf of the presumed infringed party.[22] The plaintiff can obtain from a court an injunction ordering the definitive cessation of the activities of the presumed infringer.[23]

---

[20] The Patent Reform Act (2011) recently adopted in the USA tries to include different measures addressing the insufficient incentives to challenge a weak patent, by introducing for example some adversarial procedures to challenge a patent's validity such as exist in Europe.

[21] It is interesting to note that it is precisely in this sequential innovation framework that Bessen and Maskin (2009) could show that imitation was not socially reprehensible, thus reducing the necessity for patents.

[22] The third party acting on behalf of the patent's owner is called a non-practising entity, sometimes described as a 'patent troll'.

[23] The best illustration is given by the Blackberry's story. In November 2001, the firm NTP, Inc. was filing a complaint with Virginia State Court of Justice, against the Canadian firm Research in Motion (RIM), producer of the Blackberry mobile phone, for the usage of technologies protected by old patents that were not yet expired, held by NTP. After long legal proceedings, including a re-examination of the aforementioned patents by the USPTO, the court pronounced RIM guilty,

Should the threat of injunction be considered as an abusive way for the patent holder to obtain *ex post* a substantial financial compensation in the settlement or does it correspond to the normal use of a right attached to the patent? The question is complex and does not receive a consensual answer. Some believe that a patent holder has the right to use the threat of injunction since it is a tool to deter a third party from using the patent unduly. Others believe the opposite, arguing that the threat of injunction corresponds to an opportunistic hold-up behaviour since it occurs after the victim started its productive activities.[24] The following paragraph briefly presents some contributions justifying these different positions.

6.3.1 Injunctions, hold-up and excessive royalties: Shapiro's model

The contribution by Shapiro (2010) analyses the links between the level of royalties and the injunction power in a bargaining model in which the patented innovation of an upstream producer is an input for a downstream producer. By incurring a specific cost, the downstream producer can get around the patented input and build a substitute. The model integrates the following elements: (1) the patent protecting the upstream input is uncertain, which leads to the consideration of the patent's strength as a parameter of the model; (2) the value-added derived by the downstream producer from the upstream input is assumed to be observable; (3) if a court upholds the patent's validity and recognizes the infringement, an injunction to stop the downstream production is legally executable; (4) the amount of reasonable royalties that the patent holder could have obtained *ex ante* is known by both participants; and (5) the final demand facing the downstream producer does not depend on the royalty paid for using the upstream input.

and condemned the firm to pay damages and stop production of the Blackberry. The injunction decision remained pending during the appeal procedures brought by RIM with the Court of Appeals for the Federal Circuit. After confirmation by the latter of the previous court's decision, RIM reached a private settlement in 2006, agreeing to pay $612.5 million to NTP, securing that NTP would withdraw its complaint. The most significant aspect of this example is that after the settlement, NTP's patent has been invalidated by the Patent Office!

24 The Supreme Court of the United States expressed in an ulterior affair (*eBay* v. *MercExchange*) some reservations against the justification of preliminary injunction and restricted the use of this practice to the so-called exceptional circumstances.

Ex ante, the downstream producer may be either informed or uninformed of the existence of a patent covering the input. A central investigation of the model is to assess the effect of the injunction on the royalty. This is obtained by computing the gap between the effective royalty and the reasonable royalty, the effective royalty being defined as the Nash solution of a bargaining game, in which the outcomes of a preliminary injunction serve as the threat point. Without much surprise, one of the principal results of the Shapiro model (2010) is that the gap is larger when the downstream producer decides to use the input before the uncertainty on the patent is resolved. This patent ambush effect highlights the fact that a patented input can penalize an uninformed user.[25] A more surprising result is that the excessive part of the royalty is larger when: (1) the share of the upstream input in the value-added of the final product is smaller; (2) the injunction power of the patent holder is higher; or (3) the protection of the upstream input results from a patent pending, in other words not yet published when the producer decides to incorporate the input in its product. These results led Shapiro to suggest restrictive conditions under which an injunction rule may be authorized. The exceptional conditions recommended by the Supreme Court in the *eBay* v. *MercExchange* case serve to define these restrictions: (1) the prejudice suffered by the plaintiff is irreparable; (2) monetary damages are insufficient to compensate the prejudice; (3) given the prejudices of the plaintiff and the defendant, affecting the capital of the defendant is appropriate; and (4) no component of society, other than the defendant, should be negatively affected by the injunction. Shapiro's model recommends the use by the plaintiff of a preliminary injunction only under these stringent conditions. In the following paragraph, the objections to this model are examined.[26]

6.3.2 Criticisms of Shapiro's model

The criticisms formulated by Sidak (2007) and Denicolò *et al.* (2007) are of many orders. First, they challenge a recommendation that unfairly favours the defendant to the detriment of the plaintiff. Second,

---

[25] The ability of a patent holder to negotiate *ex post* a level of royalties that is superior to what would be reasonable *ex ante* is one of the reasons advanced to explain the rate of growth of the number of patent applications.

[26] Elhauge (2008) challenges one of the conclusions of the models by Lemley and Shapiro (2007) and Shapiro (2006), which states that royalties obtained by 'surprise' are too high.

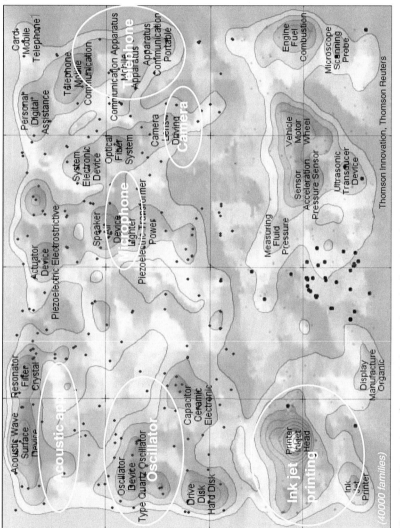

Figure 2.1 Detection of cosmetic innovations in the piezoelectric sector.

Thomson Innovation, Thomson Reuters

(40000 families)

Card-
Mobile
Telephone

Telephone
Mobile
Communication

Communication Apparatus
Mobile
Apparatus
Communication
Portable

Engine
Fuel
Combustion

Microscope
Scanning
Probe

Personal
Digital
Assistance

System
Electronic
Device

Optical
Fiber
System

Camera
Lens
Driving

Camera

Vehicle
Motor
Wheel

Piezoelectric Electrostrictive

Actuator
Device

Speaker
Oil
Device

Lighter

Piezoelectric Transformer
Power

Microphone

Sensor
Acceleration
Pressure Sensor

Ultrasonic
Transducer
Device

Measuring
Fluid
Pressure

Acoustic Wave
Surface
Device

Resonator
Filter
Crystal

coustic appl

Oscillator
Device

Type Quartz Oscillator

Oscillator

Capacitor
Ceramic
Electronic

Drive
Disk
Hard Disk

Printer
Inkjet
Head

**Ink jet
printing**

Ink
Jet
Printer

Display
Manufacture
Organic

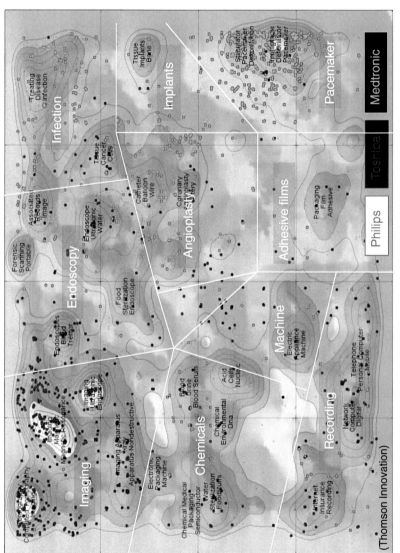

Figure 2.2 Patent mapping for medical devices.

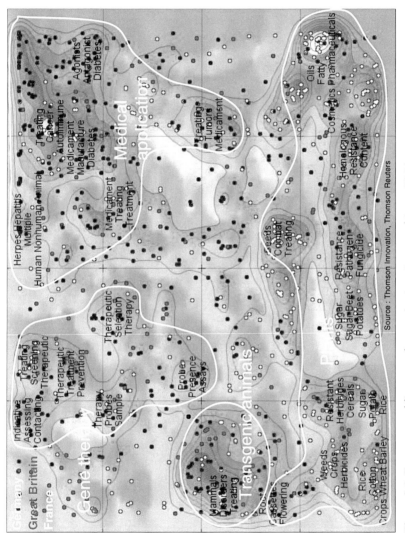

Figure 2.3 European public research organizations in recombinant DNA.

Source: Thomson Innovation, Thomson Reuters

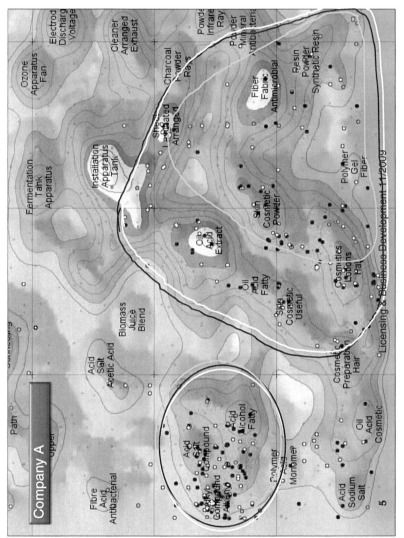

Figure 2.4 Dichroic analysis of the patent filing strategies of two companies from 1995 to 2009.

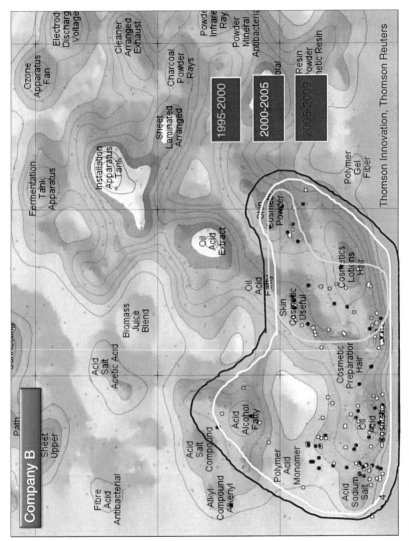

Figure 2.4 (*cont.*)

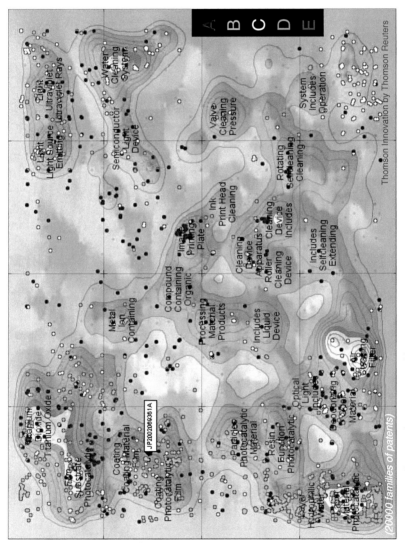

Figure 2.5 Selecting the most suitable partner.

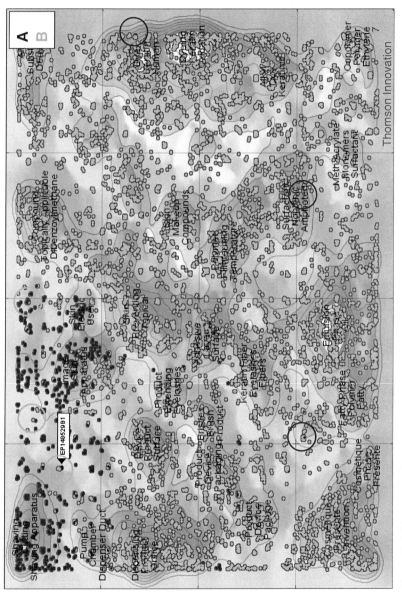

Figure 2.6 Cross-mapping.

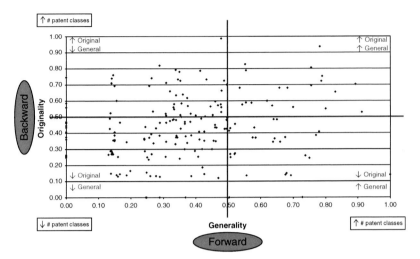

Figure 2.7 Generality and Originality Indices.

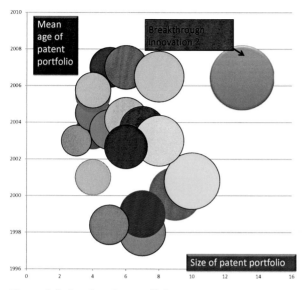

Figure 2.8 Acceleration coefficient.

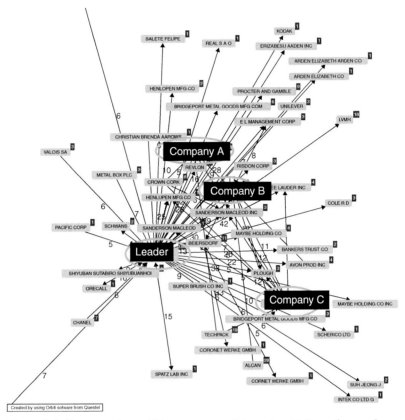

Figure 2.9 Cross-citations. (Diagram created by using Orbit software from Questel.)

they also challenge the overly negative vision of a plaintiff that is not the patent's owner. They present counter-examples of patent trolls in the biotechnology sector which had an important role in technology exchanges. Third, they show that the conclusions of the Shapiro model rest on assumptions that are not necessarily satisfied in the real world. Among the challenged assumptions are the following: (1) infringement is always detected; (2) the value added of the patented component is supposed to be small relative to the value added of the product and perfectly observable by a court; and (3) the final demand facing the downstream producer does not depend on the amount of the per-unit royalty demanded by the patent holder. The more likely assumption that information is imperfect should lead to an *ex post* level of royalty lower than the one predicted by Shapiro. Moreover, Sidak (2007) argues that since the user of the patented input benefits from the fact that the inventor had to experiment a lot before coming up with the patented innovation, he must incur costs that go well above the value added of the patented input. It is as if the user benefited from a real option value by having delayed the production of its good until the inventor of the input perfected his invention. For all these reasons, the notion of excessive royalties loses some of its relevance.

Denicolò *et al.* (2007) propose an alternative framework. The assumption is made that the value added by the patented input is private information for the defendant, and therefore is not observable by third parties, notably by a court. Thus, the amount of the reasonable royalty which serves as benchmark is itself indeterminate. It is therefore impossible for a court to evaluate the part of the royalty demanded by the plaintiff that is due solely to the injunction. The solution to this problem of asymmetric information suggested by the authors is that the court proposes an arbitrary royalty, and that the defendant uses it as an option value. If the level of the royalty proposed by the court is superior to the value of the input estimated by the defendant, the defendant has the possibility to request a renegotiation of the royalty with the plaintiff. But, if the royalty proposed by the court is inferior to the input's value, as estimated by the plaintiff, the plaintiff will have to be satisfied with this level.[27]

---

[27] The authors do not explicitly list the reasons that incite the plaintiff to accept a renegotiation if the royalty proposed by the court is superior to the value estimated by the defendant, nor the reasons that incite the plaintiff to accept the royalty proposed by the court if it is inferior to the value estimated by the defendant.

To summarize, the question of knowing whether the plaintiff should be allowed or not to use a preliminary injunction does not have a trivial answer. In any case, the answer to this question should be based on a rule of reason rather than on a per se prohibition of the injunction. In the *Apple* v. *Motorola* case, the argument of Judge Posner for refusing to ban Motorola's products from the shelves, as Apple sought, was that 'an injunction that imposes greater costs on the defendant than it confers benefits on the plaintiff reduces net social welfare'. Moreover, the case where the injunction is obtained on behalf of a patent pending must be treated with greater caution. The role of patents pending examined below is sufficiently ambiguous to justify a special treatment.

6.3.3 Role of divisional and continuation patents

According to the OECD (Zuniga and Guellec, 2009), the number of patents pending is probably greater than the volume of patents granted that are still in force. Even if a patent is enforceable only if granted, patents pending still have a value. Indeed, they offer to the depositor a temporary protection until the office has made a decision. In particular, complaints for patent infringement can be ruled in favour of the plaintiff even if the infringing behaviour took place during the time the patent was still pending. For this reason, patents pending can be instruments as powerful, if not more so, as granted patents. Furthermore, patents pending have a greater degree of uncertainty than existing patents, since it is unknown whether they will be granted, and if so on which date it will happen and whether they will correspond to strong or weak patents, since the nature and the scope of their claims are not known.

Since many divisional and continuation patent applications originate from the initial depositors, they tend to reduce the transparency of the contents of what is protected, and delay the date on which these protected claims are made public. These applications can give rise to abuses. For example, an initial depositor can claim ownership of inventions of which he is not the author, by giving as an argument the existence of a patent pending, the date of which is the initial patent date.[28]

---

[28] 'Filing of divisional applications enables some brand name pharmaceuticals to maintain the uncertainty generated by their parent patent application and to sue the producers of generic drugs that try to enter the market, in the ignorance of pending patent applications and their respective content' (European Commission, Pharmaceutical Sector Inquiry Preliminary Report, 28 November 2008).

The rule of the first to file that applies presently in all countries amplifies this risk. Some observers anticipate that the recent 'America Invents Act' law which shifted from a first to invent to a first to file principle 'will touch off a paper chase to the patent office instead of a race to innovate' (*New York Times*, 26 August 2012).

Even if it may be legal from the point of view of intellectual property law, hold-up behaviour based on pending applications constitute a large obstacle to competition, notably on technological markets, since it prevents some producers from pursuing the commercial exploitation of their activities. This situation is the reason why a report from the Federal Trade Commission in 2011 (FTC, 2011) put a particular emphasis on the requirement to clarify the notification procedures (procedures that consist in revealing the contents of what is patented and the scope of protection), both for patents granted and patents pending. The precision of the notification affects the competition at each step of the innovation process. The capacity of an operator to identify existing patents or pending applications allows him to negotiate *ex ante* a license, and thus to shield himself from the risk of apparition of blocking patents (Green and Scotchmer, 1995). In the absence of precise information, the possibility of being the victim of a threat of injunction can cause socially useful investment in R&D to never be undertaken, or lead to high levels of royalties being paid *ex post* to the patent holder. Moreover, an incomplete notification procedure does not allow potential users to negotiate *ex ante* the amount of royalties to be paid. To sum up, pending patent applications seriously affect competition conditions on the market for technological exchanges.

6.4 Patent pools, standards and competition in innovation markets

A third conception of the role of patents is based on activities that require the simultaneous use of many inputs, each being protected by an independent patent. This situation corresponds to a configuration of cumulative innovation that Scotchmer (2004) designates under the metaphor of 'research tools'. This configuration is somewhat different from the sequential innovation examined in the preceding paragraph. The central point is now the piling of scattered patents (Denicolò and Halmenschlager, 2012). This situation is common to many activities, such as the pharmaceutical industry, in which patents on genetic codes are essential for the development of new therapeutic targets, or the software industry in which a number of computer programs, which have their code protected, play a considerable role in the development of new software. The great number of patents involved impedes the prior negotiation between one user and every patent holder, giving rise to the 'tragedy of the anti-commons', a concept penned by Heller (1998) and Heller and Eisenberg (1998), and tested empirically by Murray and Stern (2007) and Cockburn, MacGarvie and Müller (2010). According to this phenomenon, fragmented property rights on a common resource lead to a sub-utilization of that resource. Furthermore, Hall and Ziedonis (2001) and Ziedonis (2004) showed that if patents on complementary innovations were highly dispersed among owners, firms requiring these patents would in turn intensively patent their own innovations to obtain a higher bargaining power with regards to the patent holders. This observation illustrates the role of patent portfolios as negotiation instruments ('bargaining chips' in the language of Hall and Ziedonis). In what follows, we examine: (1) the question of the competition bias that results from the multiplication of patents ('patent thicket') required to produce a good; (2) the problems highlighted by the grouping of these patents in a patent pool; and finally (3) the questions posed by technological standards, essential patents and their licensing rules.

6.4.1 Competition bias due to the 'patent thicket' problem

A patent thicket is a situation in which a technology is covered by many patents owned by different parties. It is defined by Shapiro (2001) in the following way: 'A dense web of overlapping intellectual property rights that a company must hack its way through in order to actually commercialize new technology.' The multiplication of patent thickets in varied sectors such as semiconductors, biotechnologies, nanotechnologies and informatics software (Ayres and Parchomovsky, 2007; IPO, 2011) poses a number of problems.[29] In particular, because of patent thickets, potential users of a technology must pay royalties to multiple patent owners, each of them benefiting to a certain extent

---

[29] The reader will find in Lemley and Shapiro (2005) an in-depth analysis of the reasons explaining the multiplication of patent thickets. Also, IPO (2011) provides an analysis of the different real-life situations that are covered by this definition.

from veto power on the new innovation. Patent thickets considerably increase the costs of negotiation between the patent owners and the user, who must obtain many licences before being able to innovate. From the viewpoint of microeconomic theory, each patent owner can be seen as a monopoly that controls an input necessary for the production of the new innovation.[30] The existence of the patent thicket could then lead to a weakening of the incentives to innovate, if the sum of the individual royalties is sufficiently high to make the planned innovation not profitable. The multiplicity of essential patents can thus block eventual innovations. This issue is important, especially when these patents are part of the elaboration of technological standards, and when users have incurred sunk costs to achieve conformity with that standard. How can we eliminate the inefficiency associated with essential patents and with the double-margin phenomenon? The economic literature proposes various solutions. A first solution consists of the establishment of a system of cross-licensing with or without reciprocal royalties. As underlined by Shapiro (2001, p. 123), 'If two patent holders are the only companies realistically capable of manufacturing products that utilize their intellectual property rights, a royalty-free cross licence is ideal from the view point of ex post competition. But any cross license is superior to a world in which the patent holders fail to cooperate, since neither could proceed with actual production and sale in that world without infringing on the other's patents.' A second solution, examined in the following paragraph, is the creation of patent pools.

6.4.2 Patent pools

Patent pools are widely used to cover technologies that are part of technological standards. We will first present arguments in favour of and against patent pools, before examining specific issues related to essential patents in technological standards.

---

[30] This problem, analogous to that of the double margin, is well known in industrial economics. An enterprise that must buy various inputs, each from separate monopolies, will pay a global price that is higher than the price that would be paid if all the inputs were sold by a single monopoly. As highlighted by Shapiro (2001), 'this is merely a magnified version of the monopoly burden resulting from the patent itself, but it is well to remember Cournot's lesson that the multiple burdens reduce both consumer welfare and the profits of patentees in comparison with a coordinated licensing approach'.

Arguments in favour of and against patent pools

A patent pool can be defined as an 'agreement among patent owners to license a set of their patents to one another or third parties' (Lerner and Tirole, 2004). Put differently, patent pools 'aim at granting a single license for a package of patents belonging to different owners' (Lévêque and Ménière, 2008). Organizing joint licensing of all the patents which read for instance on a technological standard is a way to save transaction costs that are inherent to separate licensing.[31] Examples of patent pools include MPEG-2 (1997), MPEG-4 (1998), Bluetooth (1998), DVD-ROM (1998), DVD-Video (1999), 3G Mobile Communications (2001) and One-Blue (2009). Patent pool members have long been suspected of facilitating the implementation of anti-competitive behaviour. However, this view has been challenged recently,[32] as competition authorities now recognize that they may contribute to 'integrating complementary technologies, reducing transaction costs, clearing blocking positions, decreasing infringement litigation and the uncertainties related to it, and promoting the dissemination of technologies' (US Department of Justice and the Federal Trade Commission, 2007, pp. 84–85).[33] The doctrine now defended by the US Department of Justice about patent pools is the following: (1) a pool can contain only essential patents, namely patents that are

---

[31] *'The patent pool dedicated to the MPEG-2 video compression standard is a good illustration of the efficiencies achieved by joint licensing. Owners of patents reading on MPEG-2 delegate to a jointly owned enterprise, MPEG LA, the task of licensing their patents as a single package. The pool was created in 1997 by 8 organizations holding some 100 patents representing 60% of the patents reading on the MPEG-2 standard. The pool has since expanded rapidly. In 2004 it was comprised of 650 patents owned by 25 organizations, accounting for more than 90% of the patents surrounding the standard. The MPEG-2 patent pool offers "one-stop shopping" to users of the standard. It thereby saves search and negotiation costs for would-be licenses'* (Lévêque and Ménière, 2008).

[32] Delcamp (2010), using data on 1564 essential US patents belonging to eight different pools and a control database with patents having the same characteristics, shows that patents included in a pool are more litigated than non-pool patents presenting the same characteristics (the patent holder being the plaintiff in the litigation).

[33] We do not examine in this article the questions of formation and stability of patent pools. See Aoki and Nagaoka (2004), Brenner (2009) and Lévêque and Ménière (2011). Similarly, we do not discuss the empirical literature on patent pools. See Lerner, Strojwas and Tirole (2007), Lampe and Moser (2010, 2011), Layne-Farrar and Lerner (2011), Baron and Delcamp (2010), Baron and Pohlman (2011) and Delcamp (2011).

necessary to implement a given technology;[34] (2) a pool should include a provision allowing for independent licensing (i.e. the requirement that independent licences be offered by pool members to third parties); and (3) the royalty rates and the grant-back provisions must be subject to particular monitoring.[35] The European Commission shares the positions of the US Department of Justice, but goes further in its guidelines on the structure and organization of a patent pool.[36]

Two types of questions are important in the analysis of the effects of patent pools on competition in the technological exchange market and the innovation market. The first is the degree of substitutability and/or complementarity of patents in the pool, and the second is whether patent pools should authorize patent holders to grant individual licences independently from the joint offer of the pool.

Complementary vs. substitute patents

To find out if patent pools have pro- or anticompetitive effects, we must first know whether patent pools should include only complementary patents, or if substitutable patents should be allowed as well.[37] A well-known result is that patent pools increase welfare when patents included are perfect complements, but reduce welfare when they are perfect substitutes (Gilbert, 2004; Lerner and Tirole, 2004; Lerner, Strojwas and Tirole, 2007; Shapiro, 2001). The argument is simple: a patent pool formed by patents that are perfect substitutes (i.e. distinct patents that fulfil the same functionality) is analogous to a cartel on the product market, because its goal is to eliminate competition on royalties between patent holders.[38] In the absence of

---

[34] By contrast, as underlined by Lerner and Tirole (2008), before 1995, there were almost no provisions relative to the inclusion of essential patents in the pools.

[35] Lerner and Tirole (2008, p. 167) underline that 'grant-back provisions force members of the pool to turn to the pool for free or at a low price future patents that will be deemed essential to the working of the pool'. The reader will find in that article a discussion on the introduction of grant-back provisions in the functioning rules of a pool both from the point of view of pool members and society at large.

[36] For more details, see, for example, Lerner and Tirole (2008, p. 161).

[37] The traditional definition of substitutability (respectively, complementarity) is that two goods are substitutes (complements) if increasing the price of one good raises (lowers) the demand for the other.

[38] In March 1998 the FTC challenged a pool created by Summit Technology, Inc. and Visx, Inc., on the ground that it was anticompetitive. The pool contained patents related to two different types of laser used for eye surgery, and removed

the patent pool, the situation would be analogous to Bertrand competition with license prices tending to marginal cost. The argument
according to which a pool of substitutable patents can reduce transaction costs is also invalid, because users need a license only when patents
are substitutable.[39] Conversely, when patents are perfect complements,
royalties are reduced because the pool participants internalize the
effect of their pricing on the demand for complementary patents. As a
result, a patent pool increases both profit and user welfare: 'a pool
eliminates royalty stacking and benefits both patent holders and
technology users' (Lerner and Tirole, 2008, p. 162). Two cases should
nevertheless be distinguished: a licensor vertically integrated in the
industry in which the license is used (insider), and a licensor not active
in that industry (outsider).[40]

Furthermore, it can be difficult for competition authorities to
evaluate whether patents are complements or substitutes, because
patents included in a pool are generally not pure complements or pure
substitutes. This distinction makes sense only in a dynamic perspective:
patents covering a given technology are complements when the price
for licences is low and substitutes when the price is high. Moreover,
two patents can be complements for a given functionality at a point
in time, but be substitutes for a different functionality at another time.

price competition between the two products. As the FTC stated: '*Instead of
competing with each other, the firms placed their competing patents in a patent
pool and share the proceeds each and every time a Summit or VISX laser is used*'
(Lévêque and Ménière, 2007).

[39] However, Kato (2004) shows that, under certain conditions, patent pools
composed of substitutable patents may also enhance consumer welfare.

[40] Kim (2004) shows that vertical integration always lowers the price of the final
good if there is a pool and all patents are complements. In other words, the
economic efficiency argument for patent pools is enhanced when some firms are
vertically integrated. Lévêque and Ménière (2007) also emphasize this point:
'The entry of pure patent holders has complicated the setting of cumulative
royalties within patent pools ... Pure patent owners derive their revenue solely
from licensing. Hence their interest is to leverage the market power of the pool in
order to set a high royalty. The interests of integrated manufacturers are
different because of their presence on downstream product market ... On the
one hand they are licensors who derive more revenue from a high royalty. On
the other hand, they are licensees who must pay royalties (e.g., the share of the
package royalty that is distributed to the other patent owners) for their
manufacturing activity. Because of this second effect, integrated patent owners
are more reluctant than non-integrated patent owners to charge high cumulative
royalties for the package of patents.'

Lerner and Tirole (2004) analyse the positive and negative effects on welfare of pools composed of patents that are not perfect complements or substitutes. The four main results are:[41] (1) pools with patents that are close complements have a higher probability of increasing welfare; (2) including in a pool a patent that has close substitutes outside the pool can reduce collective welfare, unless the patent pool allows its members to grant individual licenses (see the next section); (3) the incentives for patent holders to invent around or try to invalidate patents held by other members are dulled by the formation of a patent pool; and (4) pools reduce the differentiation of downstream users when the latter are also licensors.

Collective licences and/or individual licences
The pro- and anticompetitive effects of patent pools also depend on the possibility for individual owners to grant individual licenses independently of the patent pool's collective license. In the case of perfect substitutes, independent licenses would allow the re-establishment of a form of competition and so limit the monopoly power of the patent pool. However, why would the patent pool be allowed in the first place? In the case of perfect complements, independent licences make no sense. The intuition is the following: independent licences can function if users of a technology can be satisfied with the purchase of a small number of individual licences instead of buying the whole set of licences as a bundle from the patent pool. However, in this case, the problem of royalty stacking would reappear: the price paid by the users would be superior to the pool's profit-maximizing price. The immediate consequence of this royalty stacking is that the patent pool is not affected by the possibility that its members grant individual licences (Lerner and Tirole, 2008). We thus understand that the issue of independent licences makes sense only in cases in which patents are neither perfect complements nor perfect substitutes. The possibility for the licensor to license its intellectual property rights independently of the patent pool introduces a potential competition for the pool's offer, especially if the latter offers a bundle of licenses that contains more licences than what the users of a given technology need. In this

---

[41] Note that the Lerner and Tirole model (2004) makes several assumptions, some of which are strong. They are: (i) user preferences are separable; (ii) pools cannot be formed with a subset of the relevant patents; (iii) only polar cases of closed pool and pure third-party licensing by the pool are dealt with.

case, the users will prefer buying a smaller bundle at a lower price. Independent licences can thus discourage the formation of patent pools even when these pools would actually increase social welfare. Inversely, the possibility for patent holders to grant independent licenses does not always constitute a sufficient constraint for pools that eliminate competition.

Lerner and Tirole (2004, 2008a) find the following results:

> Independent licensing perfectly screens out good pools and bad pools: (i) independent licensing is an irrelevant covenant when the pool aims at lowering the overall price of the technology below the price that prevails in the absence of pooling arrangements (royalty stacking); (ii) independent licensing restores competition and re-establishes the price of the technology at the pre-pool level when the pool aims at raising the price of the technology (suppression of competition)... Overall, while the conclusion that independent licensing always screens in good pools and screens out bad ones must be qualified, the case in favor of pools with independent licensing remains quite strong, even from an ex post perspective, in our current state of knowledge (Lerner and Tirole, 2008a, pp. 166–167).[42]

Essential patents and technology standards

Whether technological standards are always beneficial for society or can lead to welfare losses is a two-sided question. On one hand, the social welfare provided by a technological standard is positive when the demand for final products that use the standard is characterized by strong network externalities and when the standard offers compatibility between the different products that satisfy the same technological norm. For example, cellular telephony would be riddled with problems if mobile phones from different makers were not mutually compatible. On the other hand, the creation of a technological standard necessitates strong cooperation between involved agents, both to obtain an

---

[42] Brenner (2009), using a set-up close to that of Lerner and Tirole (2004), addresses the issue of optimal patent pool formation when pools may be either pro- or anticompetitive, which ultimately depends on the degree of complementarity among patents. He shows that exclusive pool membership (a situation where a firm is allowed to participate in the start-up of a pool only if all other pool initiators agree) must be added to licensing rules for a pool to be welfare enhancing. Assuming that complementary patents may be either essential or non-essential, Quint (2012) finds that a pool containing only non-essential patents can reduce social welfare, even if the pool is stable to compulsory individual licensing.

agreement that will become the standard and to define conditions of its usage, notably for the license rules of essential patents included in the standard. This cooperation obviously entails a risk of collusion that can be harmful for society. To know which of these two sides finally prevail is a difficult question, and that is the reason why competition authorities, notably the Department of Justice (DoJ) and the Federal Trade Commission (FTC) in the United States, recommend an analysis based on the rule of reason and not on the application of rules per se (US Department of Justice and FTC, 2007, ch. 2).

We start by illustrating the importance of technological standards *de jure* stimulated by associations of producers and users, known under the label Standard Setting Organizations (SSO). Chiao, Lerner and Tirole (2007) have identified 59 SSOs operating in the information technology and telecommunications sectors. As underlined by Schmalensee (2009), 'as of 2001, the IEEE-SA (Institute of Electrical and Electronics Engineers Standards Association) alone had 866 active standards and 526 projects in hand, with more than 450 technical working groups and committees.' SSOs that wish to introduce a new standard must achieve a Standard Setting Agreement (SSA). The existence of SSOs can obviously create problems from the viewpoint of competition policy. Some SSOs can fall under Article 101(1) of the Treaty on the Functioning of the European Union (TFUE), in the sense that they may be considered as constituting a horizontal agreement between competitors, which is prohibited per se by Article 101(1).[43]

---

[43] The European Commission Guidelines on the applicability of Article 101 TFEU to horizontal cooperation agreements recognizes the potential beneficial impact of standard setting agreements on technological progress but warns about the potential for restrictions, too: 'Standardization agreements usually produce significant positive economic effects, for example by promoting economic interpenetration on the internal market and encouraging the development of new and improved products or markets and improved supply conditions. Standards thus normally increase competition and lower output and sales costs, benefiting economies as a whole. Standards may maintain and enhance quality, provide information and ensure interoperability and compatibility thus increasing value of the consumer (263). Standard-setting can, however, in specific circumstances, also give rise to restrictive effects on competition by potentially restricting price competition and limiting or controlling production, markets, innovation or technical development. This can occur through three channels, namely reduction in price competition, foreclosure of innovative technologies and exclusion of, or discrimination against, certain companies by prevention of effective access to the standard (264).' Note that 'a SSA may be exempted from article 101(1) TFEU by article 101(3), which exempts horizontal

An SSO can indeed distort competition in two ways: upstream between patent owners when they allow the selection of a single technology to the detriment of alternative solutions, and downstream between the users of the standard if the terms of the contract lead to licence conditions that are excessive or discriminatory. An SSA can also fall under Article 102 of the TFUE when the amount of royalties required after the standard is adopted is found excessive. Indeed, this situation can be considered as an abuse of a dominant position.[44]

Another issue highlighted by Farrell *et al.* (2007, p. 607) is that '*ex ante*, before an industry standard is chosen, there are various attractive technologies, but *ex post*, after industry participants choose a standard and take steps to implement it, alternative technologies become less attractive. Thus a patent covering a standard may confer market power *ex post* that was much weaker *ex ante*.'[45] The alternative technological solutions can even be forced to leave the market. Once a standard has been adopted, it is indeed almost impossible to replace a technological solution with another, given the costs that users would need to incur. This situation leads to a hold-up problem: 'The risk of hold-up comes from the nature of investments that manufacturers of standard-compliant equipment have to undertake for testing, designating and producing. Usually, these investments are very specific to the chosen standard. As a result, they cannot be easily redeployed to other

agreements contributing to technological progress ... In order to assess whether a standard setting agreement may be exempted from Article 101(1), one needs to establish whether an actual standard setting agreement (a) contributes to technological or economic progress, (b) while allowing consumers a fair share of the resulting benefit, (c) without imposing restrictions which are not indispensable for the attainment of the efficiencies, (d) without eliminating competition in respect of a substantial part of the products in question' (Stryszowska 2011).

[44] The *Rambus* and *Qualcomm* cases illustrate the difficulties in proving a violation of Article 102 of the TFUE. The interested reader will find in Lévêque and Ménière (2009) references for the numerous commentaries on these two cases.

[45] Interoperability is another problem in this situation. The development of a given standard can stimulate the development of another technology compatible with the new standard but incompatible with the alternative technology that was not chosen by the standard. As highlighted by Stryszowska (2011), 'for example, an introduction of a standard for a given type of microprocessor may yield a risk of the incompatibility between newly developed memory cards and an alternative microprocessor. In that case, an alternative type of microprocessor may have little chance to survive.'

uses ... Because of this lock-in effect, manufacturers are ready to pay a much higher royalty after the standard is adopted than before' (Lévêque and Ménière, 2009).

In fact, most SSOs adopted rules that allow the mitigation of the hold-up risk associated with the creation of a standard: 'These rules cluster in three areas: disclosure rules, requiring certain disclosures of patents or patent applications; negotiation rules, regarding the timing and locus of license negotiations; and licensing rules, governing the level and structure of royalties, most often requiring participants to license essential patents on "Fair, Reasonable and Non-Discriminatory" (FRAND) or "Reasonable and Non-Discriminatory" (RAND) terms' (Farrell *et al.*, 2007, p. 609).[46]

The delimitation of so-called 'essential' patents leads to decisions that are crucial in the ulterior development of the standard. However, despite the pretention that this delimitation is the result of technical expertise, it is still susceptible to two sources of bias. The first bias is due to the contents of the list of patents declared essential. Firms participating in the establishment of the standard are encouraged to identify all the patents that they consider as necessary for the usage of the standard. That list can be considerably longer than the list constituted by external experts. The first bias is thus an excess of essential patents.[47] The second source of bias has an effect opposite to the first. For strategic reasons, an enterprise can choose not to list one or many of its patents at the time of the establishment of the standard, only to use them at a later date to sue users of the standard.[48]

---

[46] Moreover, Farrell *et al.* (2007, p. 624) specify that 'the disclosure rules seek to eliminate pure hold-up and allow SSO members to judge for themselves whether other protections will adequately limit hold-up in a particular case; negotiation rules could help make negotiations better reflect *ex ante* competition, but overblown concerns about collective negotiation weaken this approach; and licensing rules are best seen as an impressive but binding default *ex ante* contract.'

[47] To illustrate, Goodman and Myers (2005) examined the patents declared essential for 3G cellular technology and made two conclusions: (1) 75% of patents (nearly 8,000) that are declared essential for that standard are owned by four companies; (2) only 21% of patents that are declared essential are truly essential, according to experts in that field including the authors of that paper.

[48] This strategy of ambush or hold-up is described by the OECD as follows: 'The implementation of an ambush strategy by means of patents pending consists of a firm not informing the standard organisation that it applied for patents in relation to the norm being established. At the same time, the firm also modifies the claims in these applications so that they correspond to the future standard.

The timing of this ambush strategy, which is often based on patents pending, is crucial, as highlighted by an OECD report (Zuniga and Guellec, 2009): 'If the enterprise had revealed the existence of its patents pending during the standard's elaboration negotiations, the standard organisation could have chosen a different, less costly technology (if possible) or could have tried to convince the firm to limit the amount of its royalties. However, keeping its patents secret until the norm is applied widely enough to prevent the establishment and application of another standard allows the firm to acquire a dominant position that it would have not acquired in other circumstances.'

Following this reasoning to the letter, we could be tempted to conclude that, in the measure that the obtainment of the dominant position is through abusive behaviour, antitrust authorities have the means to stop ambushes in technological standards organizations.[49] This task is difficult for two reasons (see Anton and Yao, 1995; Froeb and Ganglmair, 2009). The first is that antitrust authorities cannot force members of standards organizations to disclose their patents and patents pending that are related to the norm in consideration. The incentives to do so depend essentially on the internal functioning

The firm can also have an influence on the standard, so that it is closer to the claims in its patent applications. The enterprise can thus modify both the standard and its own claims in pending applications so that they coincide as much as possible. If everything goes according to plan, the standard organisation publishes a norm that is covered by the undisclosed patents pending, while the enterprise carries on with the application procedures, from the examination to the granting of the patent. Meanwhile, other enterprises apply the standard to their own products. Important irrecoverable investments are made on the basis of the norm. When the ambushing enterprise is certain that the sunk costs incurred by other firms are large enough to discourage the transition to another norm, it reveals the existence of its patents and attacks, threatening with actions to be compensated for damages. It can decide to ask for large royalties or to simply block the application of the technology in question' (Zuniga and Guellec, 2009, translated from French). This example illustrates once again the strategic role of patents pending. This strategy was used in recent cases, including those of *Rambus* and *Qualcomm*.

[49] However, note that there is a difference between Europe and the United States. In Europe, abuse of a dominant position is punished by Article 102 of the TFUE. This article is not focused solely on dominant positions acquired through abusive behaviour. Because the acquisition or the attempt to acquire a dominant position by anticompetitive manners is not within the scope of Article 102, antitrust laws of the European Treaty do not allow resort to corrective measures to combat ambush strategies in standards organizations. On the contrary, in the United States, the acquisition of a monopoly through misleading information, as is the case in the standards context, is within the scope of competition law.

rules of the standards organizations. The second reason is that antitrust authorities cannot force the organizations' members to announce the maximal price of licences and the restrictive conditions they would impose for the use of their patents. Also, the authorities do not have the power to prescribe the conditions of sale of licenses. These requirements are born more of an eventual public regulation of the market for licences than from the application of competition law. In general, standards organizations themselves set the contracting rules in these matters.

A vast literature has developed to define reasonable licence prices (see, among others, Farrell *et al.*, 2007; Gilbert, 2011a; Lévêque and Ménière, 2009). The simplest solution to the hold-up problem is to have royalties fixed before the adoption of the standard. However, this solution is difficult to implement in practice. Indeed, patent holders do not know the whole potential of the technology before the standard is developed, and thus are incapable of estimating the profits that they could enjoy from their licences. Moreover, even in the case of royalties that are a percentage of the sale price, the price elasticity of the product in the downstream market can be poorly estimated. A possible solution would be for the patent owners to commit *ex ante* not to demand excessive royalties *ex post* that result from the foreclosure of alternative technologies.

Some SSOs have thus tried to resolve the hold-up problem by asking their members to commit to having licence conditions in which *ex post* royalties are fair, reasonable and non-discriminatory (FRAND), although without specifying the exact amount *ex ante*. A violation of these terms can be subject to legal proceedings.[50] FRAND terms are

[50] Given that the notion of a fair and reasonable price is imprecise, it is not surprising that multiple controversies have emerged on the subject. Therefore, the predictability of the royalties is limited both for the licensee and the licensor. In case of litigation, courts will not be able to assess whether the amount of the royalty is fair and reasonable. Propositions have been made to better define these notions, notably by using elements of cooperative game theory. However, these definitions still face technical and legal objections, which make their application controversial. The only approach that could prevent abusive behaviour would be an *ex ante* negotiation between holders of essential patents and a user of the standard, the latter acting as a counterbalance to the seller's bargaining power. However, as mentioned above, the multiplicity of patents involved makes it difficult to hold prior negotiations between a user and all the owners of patents that are necessary to the user. Moreover, because members of the organization are themselves potential users of the norm, an *ex ante* negotiation for the conditions of sale of the licences could be interpreted as a concerted practice, revealing collusion between the members of the organization. The consequence of these difficulties is that antitrust authorities prefer adopting

obviously subject to interpretation. They can be interpreted as the licence amount that would have been negotiated *ex ante* if the economic value of the standard had been known. This interpretation supposes that the marginal contribution of each patent used in the standard can be estimated. It is obvious that royalties will be low for patents that are non-essential or replaceable and subject to *ex ante* competition, whereas royalties will be higher for patents that are essential and non-replaceable. FRAND terms are interesting precisely for replaceable patents to avoid rent extraction *ex post*, when competition has been eliminated. One proposition that is often made is to cap the amount of royalties asked by a patent holder at the incremental value of the patent to the standard (Dolmans *et al.*, 2007; Farrell *et al.*, 2007). 'This rule considers that a licence should be established according to the ex-ante incremental contribution value of the chosen technology as compared to the next best alternative, measured at the stage in which other substitute technologies were available' (Layne-Farrar, Llobet and Padilla, 2012). Swanson and Baumol (2005) propose a solution to make this proposition operational: that enterprises wishing to have their technology included in a standard agree to participate in an *ex ante* auction by offering a royalty rate. The technology of the enterprise offering the lowest royalty will be integrated in the standard. As highlighted by the authors of that proposition, 'the "best" IP option will be able to command a license fee equal to incremental cost plus the difference in value between the best and the next-best alternatives' (Swanson and Baumol, 2005, p. 23).[51] Ganglmair, Froeb and Werden (2012) propose an alternative solution in which innovators commit *ex ante* to offer an option-to-license that producers can exercise if they want to include their intellectual property in the final product. This solution is superior to the FRAND licensing system.

In this respect, the work of Layne-Farrar, Llobet and Padilla (2012) is particularly interesting. These authors start by relaxing two assumptions made implicitly in the literature to justify the incremental value rule. They challenge (1) the idea that all R&D is completed at the

a flexible interpretation of competition law towards standards organizations by encouraging them to adopt rules that limit the recourse to ambush strategies. When they intervene directly in cases of abuse leading to dominant positions, antitrust authorities use the rule of reason, and not prohibition per se.

[51] Incremental cost is defined as the cost per unit that patent holders incur as a consequence of licensing their patent.

moment of the adoption of the standard and that the innovations were available to be used in the standard, and (2) that enterprises holding patents all choose to join an SSO. The authors show that if patent holders have the choice (and not the obligation) to join an SSO, then they generally prefer to stay out of the SSO even if the latter imposes an incremental value rule.[52] The intuition is the following: the best strategy for patent holders is to stay out of the SSO and hope that the substitutable technology chosen by the default is non-functioning.[53] This strategy would allow these outsiders to find themselves in a position of strength *ex post* and to ask for higher royalties. The authors also find that 'in order to ensure the patent holder's participation, SSO members are able to and will be interested in increasing the licensing fees paid to the patent holder above the dictated level by the incremental value'. More precisely, these results depend on the degree of complexity of the standards. In the case of simple standards formed of a small number of complementary components, R&D efforts of participating enterprises are complementary; thus if an enterprise chooses to join an SSO, its participation will provoke an increase in the R&D of other firms. This chain of events increases the standard's probability of success and thus the probability to see patent holders participate in the standard despite the licensing cap introduced by the incremental value rule. However, this participation is less important in the case of complex standards, because complementarities are dispersed over a greater number of firms. Finally, the authors conclude that 'the overall picture indicates that licensing caps like the incremental value rule tend to reduce the incentives for firms to participate in cooperative standards setting since most standards can be qualified as complex'.

6.5 Conclusion

In this chapter, we have shown that the contemporary tensions between intellectual property and competition, other than the classical trade-off between static and dynamic efficiency, originate in the

---

[52] Note that this result is not specific to incremental licensing rules and arises for other licensing rules like the benchmark licensing proposed by Lemley and Shapiro (2007).

[53] The authors assume 'that the patented technology is superior to the default one but the standard might still be viable if patent holders do not participate, and the SSO needs to rely on the default technology'.

evolution of the patent system. Some characteristics of this system indeed lead to distortions in competition that cannot be corrected by a simple application of antitrust laws. Three of these characteristics were examined in details.

(1) Many patents of dubious validity are approved by patent offices
Patents represent the sacrifice that society is willing to make to benefit from the introduction of a new product or production process. This sacrifice is manifested by a temporary exclusion power granted to the patent holder. For this sacrifice to make sense, patentability criteria must be fully satisfied, and the scope of the claims protected by the patent must not be excessive. However, for a variety of reasons examined in this chapter, many patents granted by patent offices do not satisfy these criteria, even when they cover products of great social value. Consequently, competition on the products market is affected: the temporary exclusion power is no longer justified even if holders of dubious patents try to rely on it. It is not the role of competition authorities to verify whether practices of patent holders rely on dubious patents or not. However, the increased uncertainty on the quality of existing patents implies that the social costs of patents are increasing. This cost is measured in two ways: either by the legal cost of conflict resolution if the litigation is taken to court, or by the social cost induced by a settlement that can still lead to a welfare loss for consumers. The efficiency of the patent attribution system is a necessary and essential prior for competition law to prevent an abusive usage of the patent, in other words a usage that goes over the legal privileges attached to the title.

(2) The usage of patents in sequential innovations enables a hold-up behaviour that affects competition on technological markets
Beyond simple intellectual protection against the risk of copying, patents increasingly fulfil an essential role on technological markets as a medium of exchange. Consequently, the protection of an innovation against imitation is often accompanied by a potential blocking power against ulterior innovations that require the use of patented technologies. This blocking power is particularly detrimental, because it is often exercised *ex post*, after the second innovator has made the investments necessary to his activity. Technological exchanges can thus be constrained in multiple ways, following a licensing rejection, an excessive usage fee, the usage of a

patent pending, or the injunction power that the patent holder can use to stop the activity of the potential infringer. The powers of competition authorities seem limited when dealing with excessive protection, and the usual presumption that a reinforcement of intellectual property always favours innovation is not verified in the case of sequential innovation. Reducing the abusive hold-up behaviour of the holder of a dubious patent is a task that rests upon courts and regulation agencies, and only in part on competition authorities.

(3) The simultaneous usage of many patented inputs in complex technologies imposes limits to the competition for innovation
Another issue beyond the sequential conception to innovation is the effects of fragmented ownership of intellectual property and the quantity of patents necessary for the production of some goods. On one hand, the patent thicket necessary to access the market can be an obstacle for competition. The constitution of patent pools with a collective license can sometimes mitigate this problem, but does not solve other questions such as the nature of, and interactions between the patents constituting the pool, and the degree of autonomy in the granting of licences. On the other hand, the implementation and functioning of a technological standard necessitates close collaboration mechanisms between the holders of essential patents. This coordination can cause problems in the smooth operation of the innovation market. In particular, the establishment of general principles that must be satisfied by the collective licence's price does not suffice to mitigate the risks of individual deviation and hold-up. Many highly publicized trials illustrate well the materiality of opportunistic behaviour in the functioning of a standard. However, it is again difficult for competition authorities to prevent these risks or to convict the authors of opportunistic behaviour.

To summarize, the argument defended in this chapter is that we have arrived at a new stage of tensions between intellectual property and competition.

In the previous stage, the goal was to draw borders between intellectual property law and competition law. This task was difficult but feasible. After successive phases, characterized by the alternative dominance of one law over the other,[54] of which we can observe the

---

[54] At the beginning of the twentieth century, patent law seemed to dominate competition law, so that clauses restricting competition in licence contracts were

traces in the first guidelines established by competition authorities and in the court decisions, a general principle emerged, consisting of the recognition of the specificity of intellectual property before applying antitrust rules. For example, to know whether a restrictive clause in a licence contract violates competition law or not, authorities do not judge the restriction on competition per se as if it was applied to a standard market activity. Authorities instead compare the level of competition with the restrictive clauses in the licence to the hypothetical level of competition that would prevail if the license did not exist. If it is estimated that competition would be stronger in the absence of the license, the restrictive clauses are judged as violating competition law. However, if competition would not be stronger without the licence, the restrictive clause is not judged as anticompetitive by antitrust authorities. In cases at the border of competition and intellectual property, the application of the reason rule thus progressively replaced per se rules.

In the current stage, the origin of the conflict between patents and competition seem to be of a different nature. The conflict resides in the fact that patents are of uncertain quality, and that there are close links between fragmented intellectual property rights. The granting of patents, and the wars that sometimes accompany these decisions, are becoming decisive episodes in the battles between producers operating with complex technologies. To illustrate, we can take a recent example of a transaction involving many patents. Microsoft decided to sell to Facebook more than 800 patents related to advertising, research, online commerce and mobile telephony for over $1 billion, not long after Microsoft had bought these patents from AOL for $550 million ('L'Usine nouvelle', Reuters, 24 April 2012). How can we analyse this type of transaction? Is it a simple transfer of immaterial assets, translating the existence of comparative advantages according to which Facebook would be better suited than AOL in the exploitation of these patents, while Microsoft only plays the role of a lucrative middleman? Or is it instead a reinforcement of the economic power of Facebook due to the acquisition of a set of patents that play a considerable role

generally allowed. A reversal happened in the 1960s, when competition law started to dominate over patent law. Prohibition criteria were thus applied to some competition-restricting clauses, known in the United States as the '9 no-no's'. For a more detailed history of the evolution of the relations between intellectual property law and competition law, see Gilbert (2006).

in the negotiations between actors in the digital economy? What is the legal validity of these patents, and how will eventual legal proceedings concerning some of them be resolved? These questions are complex and illustrate the diversity of the potential conflicts between patents and competition. Another illustration is the recent case before a court in San José in California, in which Apple sued Samsung for infringement of several patents used in mobile phones and tablet computers. Because the case was heard before a popular jury that did not necessarily have all the technical knowledge pertaining to the case, the arguments of the plaintiff (Apple) as well as the arguments of the defendant (Samsung) rested on abstract principles. For the plaintiff, the case was not a war between two giants, but simply an opportunity to save the principles and values of intellectual property: Samsung allegedly imitated in three months what Apple developed in three years. For Samsung's lawyer, the principle of competition was violated:

Today's verdict should not be viewed as a win for Apple, but as a loss for the American consumer. It will lead to fewer choices, less innovation, and potentially higher prices. It is unfortunate that patent law can be manipulated to give one company a monopoly over rectangles with rounded corners, or technology that is being improved every day by Samsung and other companies. Consumers have the right to choices, and they know what they are buying when they purchase Samsung products. This is not the final word in this case or in battles being waged in courts and tribunals around the world, some of which have already rejected many of Apple's claims. Samsung will continue to innovate and offer choices for the consumer.

The Economist analysed the conflict between Apple and Samsung and stated that: 'If, as it seems, Apple has had to resort to the courts to stifle competition and limit consumer choice, then it is a sad day for American innovation. That the company can do so with such impunity is an even sadder reflection of how dysfunctional the patent system in the United States has become.' This battle of principles is at the heart of contemporary tensions between intellectual property and competition. Furthermore, the verdict of the American court that Samsung had infringed on Apple's patents was not the same in a judgment in front of a Japanese court.[55] Should we see this difference in the judgments as a sign that the United States recognizes the applicability of patents to computer software, while other countries have some reservations?

[55] The Japanese judgment involved a slightly different set of patents.

As a last word, the evolution of technology invites us to revisit the foundations of the patent system to analyse the new relations between competition and intellectual property, which we have tried to sketch in this chapter.

References

Aghion, P. and R. Griffith (2005) *Competition and Growth: Reconciling Theory and Evidence*. Cambridge, MA: MIT Press.

Aghion, P, C. Harris, P. Howitt and J. Vickers (2001) Competition, imitation and growth with step-by-step innovation. *Review of Economic Studies*, **68**, 467–492.

Allison, J. and M. Lemley (1998) Empirical evidence on the validity of litigated patents. *AIPLA Quarterly Journal*, **26**, 185–271.

Amir, R., D. Encaoua and Y. Lefouili (2011) *Per-Unit Royalty vs. Fixed Fee: The Case of Weak Patents*, Working Paper. Paris: Université Paris I Panthéon-Sorbonne.

Anton, J. and D. Yao (1995) Start-ups, spin-offs, and internal projects. *Journal of Law, Economics and Organization*, **11**, 362–378.

———— (2004) Little patents and big secrets: managing intellectual property. *RAND Journal of Economics*, **35**, 1–22.

Aoki, R. and S. Nagaoka (2004) *The Consortium Standard and Patent Pools*, Discussion Paper 32. Tokyo: Hitotsubashi University.

Appelt, S. (2010) *Authorized Generic Entry Prior to Patent Expiry: Reassessing Incentives for Independent Generic Entry*, Discussion Paper 2010–23. Munich, Germany: Department of Economics, University of Munich.

Arundel, A. and I. Kabla (1998) What percentage of innovations are patented? Empirical estimates for European firms. *Research Policy*, **27**, 127–141.

Ayres, I. and P. Klemperer (1999) Limiting patentees' market power without reducing innovation incentives: the perverse benefits of uncertainty and non-injunctive remedies, *Michigan Law Review*, **97**, 985–999.

Ayres, I. and G. Parchomovsky (2007) *Tradable Patent Rights: A New Approach to Innovation*, Research Paper 07–23. Philadelphia, PA: University of Pennsylvania, Institute for Law and Economics.

Baron, J. and H. Delcamp (2010) *Strategic Inputs into Patent Pools*, MINES Paris Tech Working Paper 05. Paris: CERNA.

Baron, J. and T. Pohlman (2011) *Patent Pools and Patent Inflation*. http://ftp.zew.de/pub/zew-docs/veranstaltungen/innovationpatenting2011/papers/

Bessen, J. and E. Maskin (2009) Sequential innovation, patents, and imitation. *RAND Journal of Economics*, **40**, 611–635.

Bessen, J. and S. Meurer (2008) *Patent Failure: How Judges, Bureaucrats and Lawyers Put Innovation at Risk*. Princeton, NJ: Princeton University Press.

Boldrin, M. and D. Levine (2008) *Against Intellectual Monopoly*. Cambridge: Cambridge University Press.

Brenner, S. (2009) Optimal formation rules for patent pools. *Economic Theory*, **40**, 373–388.

Chiao, B., J. Lerner and J. Tirole (2007) The rules of standard setting organizations: an empirical analysis. *RAND Journal of Economics*, **38**, 905–930.

Cockburn, I., M. MacGarvie and E. Müller (2010) Patent thickets, licensing and innovative performance. *Industrial and Corporate Change*, **19**, 899–925.

Cohen, W., R. Nelson and J. Walsh (2000) *Protecting their Intellectual Assets: Appropriability Conditions and Why US Manufacturing Firms Patent (or Not)*, NBER Working Paper 7552. Cambridge, MA: National Bureau for Economic Research.

Delcamp, H. (2010) *Are Patent Pools a Way to Help Patent Owners Enforcing Their Rights?* MINES Paris Tech Working Paper. Paris: CERNA.

(2011) *Essential Patents in Pools: Is Value Intrinsic or Induced?* MINES Paris Tech Working Paper 04. Paris: CERNA

Denicolò, V. and C. Halmenschlager (2012) Optimal patentability requirements with fragmented intellectual property rights. *European Economic Review*, **56**, 190–204.

Denicolò, V., D. Gerardin, A. Layne-Farrar and A. Jorge Padilla (2007) *Revisiting Injunctive Relief: Interpreting eBay in High-Tech Industries with Non-Practicing Patent Holders*. http://ssrn.com/abstract=1019611

Dolmans, M., R. O'Donoghue and P. J. Loewenthal (2007) Article 82 EC and intellectual property: the state of the law pending the judgment in *Microsoft v. Commission*. *Competition Policy International*, **3**, 107–144.

Elhauge, E. (2008) Do patent holdup and royalty stacking lead to systematical excessive royalties? *Journal of Competition Law & Economics*, **4**, 535–570.

Encaoua, D. and Y. Lefouili (2005) Choosing intellectual protection: imitation, patent strength and licensing. *Annales d'Economie et de Statistique*, **79–80**, 241–271. (Reprinted in *Contributions in Memory of Zvi Griliches* (2010), eds. J. Mairesse and M. Trajtenberg, pp. 241–271. Cambridge, MA: National Bureau of Economic Research.)

Encaoua, D. and D. Ulph (2005) *Catching up or Leapfrogging: The Effects of Competition on Innovation and Growth*, Working Paper. Paris: Université Paris 1 Panthéon-Sorbonne.

Encaoua, D., D. Guellec and C. Martinez (2006) Patent systems for encouraging innovation: lessons from economic analysis, *Research Policy*, **35**, 1423–1440.

European Commission (2008) *Pharmaceutical Sector Inquiry Preliminary Report*. Brussels: Directorate-General of Competition.

Farrell, J., J. Hayes, C. Shapiro and T. Sullivan (2007) Standard setting, patents, and hold-up. *Antitrust Law Journal*, 3, 603–669.

Federal Trade Commission (2003) *To Promote Innovation: The Proper Balance of Competition and Patent Law and Policy, A Report*. Washington, DC: DoJ and FTC.

(2011) *The Evolving IP Marketplace: Aligning Patent Notice and Remedies with Competition, A Report*. Washington, DC: FTC.

Froeb, L. and B. Ganglmair (2009) *An Equilibrium Analysis of Antitrust as a Solution to the Problem of Patent Hold-Up*, mimeo, University of Zurich and Vanderbilt University, Nashville, TN.

Ganglmair, R., L.M. Froeb and G.J. Werden (2012) Patent hold-up and antitrust: how a well-intentioned rule could retard innovation. *Journal of Industrial Economics*, 60, 249–273.

Gilbert, R. (2006) Competition and innovation. *Journal of Industrial Organization Education* 1.1. http://works.bepress.com/richard_gilbert/15

(2011a) A world without intellectual property? A review of Michele Boldrin and David Levine's *Against Intellectual Monopoly*. *Journal of Economic Literature*, 49, 421–432.

(2011b) *Deal or No Deal? Licensing Negotiations by Standard Setting Organizations*, University of California Working Paper. http://works.bepress.com/richard_gilbert/wp

(2004) Antitrust for patent pools: a century of policy evolution. *Stanford Technology Law Review*, 3, http://works.bepress.com/richard_gilbert/11

Goodman, D. and R. Myers (2005) 3G cellular standards and patents. *IEEE WirelessCom*, 13 June.

Graham, S., B. Hall, D. Harhoff and D. Mowery (2004) Patent quality control: a comparison of US patent re-examinations and European patent oppositions. In *Patents in the Knowledge-Based Economy*, eds. W. Cohen and S. Merrill. Washington, DC: National Academies Press.

Green, J. and S. Scotchmer (1995) On the division of profit in sequential innovation. *RAND Journal of Economics*, 26, 20–33.

Guellec, D. and B. van Pottelsberghe de la Potterie (2007) *The Economics of the European Patent System*. Oxford: Oxford University Press.

Hall, B., C. Helmers, M. Rogers and V. Sena (2012) *The Choice between Formal and Informal Intellectual Property: A Literature Review*, NBER Working Paper 17983. Cambridge, MA: National Bureau of Economic Research.

Hall, B. and R. Ziedonis (2001) The patent paradox revisited: an empirical study of patenting in the US semiconductor industry, 1979–1995. *RAND Journal of Economics*, 32, 101–128.

Harhoff, D. and M. Reitzig (2004) Determinants of opposition against EPO patent grants: the case of biotechnology and pharmaceuticals. *International Journal of Industrial Organization*, 22, 443–480.

Harhoff, D., B. Hall, G. von Graevenitz, K. Hoisl and S. Wagner (2007) *The Strategic Use of Patents and Its Implications for Enterprise and Competition Policies*. Brussels: European Commission.

Hegde, D., D. Mowery and S. Graham (2007) *Pioneers, Submariners or Thicket Builders: Which Firms Use Continuations in Patenting*, NBER Working Paper 13153. Cambridge, MA: National Bureau of Economic Research.

Heller, M.A. (1998) The tragedy of the anticommons: property in the transition from Marx to markets. *Harvard Law Review*, 111, 621–688.

Heller, M. and R. Eisenberg (1998) Can patents deter innovation? The anticommons in biomedical research. *Science*, 280, 698–701.

Intellectual Property Office (2011) *Patent Thickets: An Overview*. Newport, UK: IPO.

Jaffe, A.B. and J. Lerner (2004) *Innovation and Its Discontents: How our Broken Patent System Is Endangering Innovation and Progress, and What to Do about It*. Princeton, NJ: Princeton University Press

Kato, A. (2004) Patent pool enhances market competition. *International Review of Law and Economic*, 24, 255–268.

Kim, S.-H. (2004) Vertical structure and patent pools. *Review of Industrial Organization*, 25, 231–250.

Lampe, R. and Moser, P. (2010) Do patent pools encourage innovation? Evidence from the nineteenth-century sewing machine industry. *Journal of Economic History*, 70, 898–920.

(2011) *Patent Pools and the Direction of Innovation: Evidence from the Nineteenth-Century Sewing Machine Industry*, NBER Working Paper 17573. Cambridge, MA: National Bureau of Economic Research.

Langinier, C. and P. Marcoule (2009) *Search of Prior Art and Revelation of Information by Patent Applicants*, Working Paper 2009–21. Department of Economics, University of Alberta, Edmonton, Canada.

Layne-Farrar, A. and J. Lerner (2011) To join or not to join: examining patent pool participation and rent sharing rules. *International Journal of Industrial Organization*, 29, 294–303.

Layne-Farrar, A., G. Llobet and J. Padilla (2012) *Payments and Participation: The Incentives to Join Cooperative Standard Setting Efforts*, www.cemfi.es/~llobet/joinSSO.pdf

Lei, Z. and B. Wright (2010) *Why Weak Patents? Rational Ignorance or Pro-Customer Tilt*. Berkeley, CA: Department of Agricultural and Resource Economics, University of California–Berkeley.

Lemley, M. (2001) Rational ignorance at the patent office. *Northwestern University Law Review*, **95**, 1–35.

(2012) *Fixing the Patent Office*, NBER Working Paper 18081. Cambridge, MA: National Bureau of Economic Research.

Lemley, M. and C. Shapiro (2005) Probabilistic patents. *Journal of Economic Perspectives*, **19**, 75–98.

(2007) Patent holdup and royalty stacking. *Texas Law Review*, **85**, 1–59.

Lerner, J., M. Strojwas and J. Tirole (2007) The design of patent pools. *RAND Journal of Economics*, **38**, 610–625.

Lerner, J. and J. Tirole (2004) Efficient patent pools. *American Economic Review*, **94**, 691–711.

(2008a) Public policy toward patent pools. In *Innovation Policy and the Economy*, vol. 8, eds. A. Jaffe, J. Lerner and S. Stern. Chicago, IL: University of Chicago Press.

(2008b) *Does Independent Licensing Cure Bad Agreements?* Harvard University and IDEI–University of Toulouse, France.

Lerner, J., M. Strojwas and J. Tirole (2003) *The Structure and Performance of Patent Pools: Empirical Evidence*, Working Paper, IDEI–University of Toulouse, France.

Lévêque, F. and Y. Ménière (2008) Technology standards, patents and antitrust. *Competition and Regulation in Network Industries*, **9**, 29–41.

(2009) *Vagueness in RAND Licensing Obligations Is Unreasonable for Patent Owners*, Working Paper 2009–04. Paris: CERNA.

(2011) Patent pool formation: timing matters. *Information Economics and Policy*, **23**, 243–251.

Levin, R., A. Klevorick, R. Nelson and S. Winter (1987) Appropriating the returns from industrial research and development. *Brookings Papers of Economic Activities*, **3**, 783–831.

Murray, F. and S. Stern (2007) Do formal intellectual property rights hinder the free flow of scientific knowledge? An empirical test of the anti-commons hypothesis. *Journal of Economic Behavior and Organization*, **63**, 648–687.

Quint, D. (2012) *Pooling with Essential and Nonessential Patents*, www.ssc.wisc.edu/~dquint/papers/quint-patent-pools.pdf

Schmalensee, R. (2009) Standard-setting, innovation specialists and competition policy. *Journal of Industrial Economics*, **17**, 526–552.

Scotchmer, S. (2004) *Innovation and Incentives*. Cambridge, MA: MIT Press.

Shapiro, C. (2001) Navigating the patent thicket: cross licenses, patent pools, and standard setting. In *Innovation Policy and the Economy*, vol. 1, eds. A. Jaffe, J. Lerner and S. Stern. Cambridge, MA: MIT Press.

(2003) Antitrust limits to patent settlements. *RAND Journal of Economics*, **34**, 391–411.

(2006) Prior user rights. *American Economic Review*, **96**, 92–96.

(2010) Injunctions, hold-up, and patent royalties. *American Law and Economics Review*, **12**, 509–557.

Sidak, G.J. (2007) Holdup, royalty stacking, and the presumption of injunctive relief for patent infringement: a reply to Lemley and Shapiro. *Minnesota Law Review*, **92**, 714–748.

Stryszowska, M. (2011) EU competition law vs. standard setting agreements. *Global Antitrust Review*, www.microeconomix.eu/publications/

Swanson, D. and W. Baumol (2005) Reasonable and nondiscriminatory (RAND) royalties, standards selection, and control of market power. *Antitrust Law Journal*, **73**, 1–58.

US Department of Justice and Federal Trade Commission (2007) *Antitrust Enforcement and Intellectual Property Rights: Promoting Innovation and Competition*. Washington, DC: FTC and DoJ.

Van Pottelsberghe de la Potterie, B. (2010) *The Quality Factor in Patent Systems*, Working Paper 027. Brussels: European Center for Advanced Research in Economics and Statistics.

Ziedonis, R. (2004) Don't fence me in: fragmented markets for technology and the patent acquisition strategies of firms. *Management Science*, **50**, 804–820.

Zuniga, M. and D. Guellec (2009) *Who Licenses Out Patents and Why? Lessons from a Business Survey*, STI Working Paper 2009/5. Paris: Organisation for Economic Co-ordination and Development.

7 | Valorization of public research results and patents: elements of international comparison

RÉMI LALLEMENT

7.1 Introduction

Given that innovation and the knowledge-based economy are key dimensions to improve welfare and international competitiveness, the governments of most industrialized countries as well as of major emerging countries quite rightly invest large amounts of public money in the domain of research, notably in universities and other publicly funded research institutions. Taxpayers' money nonetheless has to be spent wisely and appropriately, all the more so in an overall context of budgetary restraint. Considering this, policy-makers, scholars and business representatives in many countries express a growing concern about the capacity of these public research organizations (PROs) to contribute effectively to wealth creation. In this respect and in socio-economic terms, the performance of innovation systems undeniably depends not only on scientific and technological excellence on the part of PROs – reflected by indicators such as the number of publications or patentable inventions – but also on the ability to build rapidly enough on those cognitive resources, transforming them into new processes, new products, new organizational forms, and ultimately into value added and job creation. This is the main issue at stake behind the notion of valorization of public research results.

Beyond this broad and relatively consensual idea of valorization, a series of points remain more controversial. In this perspective, the present chapter aims at contributing to clarify several points about commercialization (or monetization) of public research, and to dispel a series of common misunderstandings regarding this matter and about the role played by patents in this domain. Is the valorization process only a one-way technology transfer from the academy to the business sector, with PROs aiming at getting a high return on investment? Or, more broadly, is it about promoting various kinds of science–industry links, including in the form of sponsored or collaborative research, and not necessarily with an attempt to maximize financial income? What are the major channels through which this process can be performed? What are the appropriate criteria to assess the performance of the technology transfer offices (TTOs)? Should patents be considered as inputs or as outputs, in this regard? What does the evidence say about the respective positions of the main developed countries with respect to such indicators? To what extent is the USA performing better than comparable countries, according to the relevant metrics? Should other countries try to catch up by imitating the policy adopted in the USA?

It is true that an important milestone in this debate is the Patent and Trademark Law Amendments Act enacted in the USA in 1980, more commonly known as the Bayh–Dole Act, which deals with intellectual property arising from federal government-funded research, and has been quite influential on subsequent reforms carried out in other countries. If most experts approve the underlying philosophy of this law, many criticize some of its unintended effects – notably on the direction and ethics of academic research – and are at the least sceptical about the need and the possibility to replicate the Bayh–Dole Act in other countries. Though, even if national reforms have to be attuned to idiosyncratic elements of the respective countries, there are good reasons to think that lessons emanating from specific national contexts can to some extent be learned abroad.

Against this background, this chapter examines successively four main aspects of the topic. Firstly, it analyses the notion of valorization of public research results, advocating for a broad approach encompassing also the newer dimensions of 'open collaborative innovation'. Secondly, it emphasizes the difficulty of comparing and interpreting the available performance indicators, at the micro level of the respective TTOs. Thirdly, it explains how performance gaps between individual PROs or whole countries are largely due to structural and institutional factors, so that the landscape in the domain of valorization is shaped by a relatively high degree of diversity worldwide. Fourthly, it shows that, beyond this international variety of practices, it is possible to identify not only common factors for the success of TTOs, but also a general need to clarify the ultimate goals of valorization, as well as a common necessity to acknowledge and limit possible negative impacts, in case of excessive commercialism. The chapter concludes by suggesting that alternative approaches often deserve to be explored.

7.2 The necessity to go beyond a too narrow approach of valorization

As a tentative approach, valorisation of public research can be defined as an activity that increases the economic value of the outputs resulting from public research. Beyond this preliminary and vague definition, the different conceptions related to valorization can be classified in two categories.

7.2.1 *A narrow approach, based on a conception of innovation as a linear process*

The first one is a narrow approach, in which valorization (or commercialization) is synonymous for one-way technology transfer. Given that basic research is mainly done in publicly funded research institutions, the results of this research have to be translated toward the business sector. Starting from this correct assumption, this approach corresponds to a conception of technological innovation as a linear and sequential process, where ideas stem from basic research, are transformed into applied research, and then go through a phase of development leading to patent filing, licences, eventually to spin-off creation (start-up activities), and at the end result in new production processes or new goods or services introduced on the market. Intellectual property rights (IPRs) and particularly patents thus play a key and unambiguous role in this vision of valorization: they protect the transfer of publicly funded inventions to private firms willing to invest in their commercialization, and ultimately they are supposed to generate a substantial stream of royalty income for the PROs concerned.

In this perspective, where innovation is largely conceived as a process of science and technology push, the knowledge or technology transfer offices (KTOs or TTOs) try to solve a classic problem of information asymmetry: the public laboratories generally have difficulties in evaluating in advance the commercial potential of their inventions, while the firms usually have a limited capacity to assess *ex ante* the quality of these inventions (Debackere, 2012). This dimension of valorization and this conception of TTOs acting as technology brokers between public laboratories and private businesses are undeniably important, but this traditional and often simplistic vision of innovation and technology transfer is challenged.

7.2.2 A broader approach encompassing also the newer dimensions of 'open innovation'

One reason for it is that innovation is in most cases more a non-linear and interactive process with frequent retroactions, as described in the chain-linked model of Kline and Rosenberg (1986). This circular and iterative conception of innovation is more in accordance with the modern world, where most pre-competitive research is still performed by PROs, but where science and business are increasingly intertwined. The notion of 'open innovation' accounts for this frequent phenomenon, where innovation processes involve collaboration with third parties (contract and consortium research). In such cases, whole segments of private firms' R&D activities (as well as specific training activities or consultant services) are sometimes outsourced vis-à-vis PROs. And the research programmes of the PROs can themselves be inspired to a large extent by the needs of industry, customers and society, at different stages of the innovation process, so that innovation is nowadays often conceived as a market-pulled (or user-driven) process. In France, the CEA (Atomic Energy Commission) illustrates this logic, as sponsored research – i.e. R&D activities performed for (public or private) third parties – belongs to the core missions of this PRO. In Germany to a large extent, the same applies to the Fraunhofer Institutes, for which contract research generates more than 80% of the total research budget. In today's ecosystems of innovation, indeed, the interaction between industry and science does not only take place *once* the public financed research has produced technological results, but often takes place also and to a large extent *during* the process of the technological 'production'. As stressed in a Japanese study, what is labelled as 'university technology transfer' is actually more often a creation process than a transferring process (Senoo *et al.*, 2009).

This means that the notion of public research *results* is not obvious, and that an important dimension of valorization actually can also occur in the upstream phase of the innovation process, before the technology is patented or even conceived. Intellectual property rights then play a more subtle role, enabling the public and private partners to implement a deeper division of labour in this domain of collaborative research, and to coordinate their decisions accordingly. In this context, moreover, patents are not only an output for the PROs concerned, a passive reflection of their research activity and a sort of raw material waiting to be transformed by firms into innovative products or processes, because they can also have other functions. For the academic inventor and his laboratory, patents can for instance be used as a way to maintain a more or less exclusive control over subsequent research and development (Kesan, 2009), or as a signal to show a technological capacity to collaborate with industrial partners (Fritsch *et al.*, 2007).

In sum, this broader approach of valorization leads to emphasis on the need to take account of the various existing links between public laboratories and companies. While the narrow conception only considers the downstream aspect – i.e. the transfer of already existing technological outputs – the broader conception also integrates the upstream dimension of collaborative R&D. In this context, the KTOs in charge of valorization have to deal with a broad spectrum of tasks, including not only patent filing, licensing and spin-off creation, but also cooperation management, at different stages of the innovation process.

7.3 The difficulty of comparing and interpreting the available performance indicators at the micro level

This diversity of tasks explains why performance assessment, in the domain of valorization, has to take into account a series of indicators. Moreover, the complexity of international comparison in this domain is increased by the relative scarcity of available data.[1] Anyway, these

---

[1] In this regard, the yearly surveys conducted by the Association of University Technology Managers (AUTM) since 1991 provide undoubtedly the best data set for North America, with each year between 120 and 160 respondents in the USA and around 30 respondents in Canada. The data for Europe are usually less regular, less coherent and less comprehensive. The corresponding association in Europe, ProTon Europe, has collected data only since 2003 and its eighth annual report referring to fiscal year 2010 has a high number of respondents (295 PROs) but only covers six countries: UK, Italy, Spain, Denmark, Ireland and Belgium (Piccaluga *et al.*, 2011). Another interesting survey for Europe was conducted by the CEMI (Chair of Economics and Management of Innovation, Ecole Polytechnique Fédérale de Lausanne): it is more representative and features data for 211 respondents covering 15 West European countries, but only for a single year – 2007 (Conti and Gaule, 2011). As for the surveys conducted on the technology transfer activities of ASTP (Association of European Science and Technology Transfer Professionals) members, they are based on a smaller number of respondents (universities and other PROs) but cover more than 20 European countries (Arundel and Bordoy, 2008, 2010). More recently, a report prepared on behalf of the European Commission features results of the first European Knowledge Transfer Indicators Survey (EKTIS), which was conducted in 2011,

data can be classified in two categories: a first group comprises indicators about the *commercial potential* addressed by the respective TTOs (invention disclosures, patent applications, patent grants, etc.), while the second group highlights the *actual use* of this portfolio (spin-offs, licences, licensing revenue). This distinction between commercial potential and actual use is based on Arundel and Bordoy (2008). The underlying idea is that indicators such as patent numbers represent certain results (outputs) of academia concerning the creation of codified technological knowledge, but should be distinguished from other indicators such as the number of licence deals or the number of spin-offs created, which reflect more the outcome of the valorization process, in relation to users (industrial partners) and leading to wealth creation. To account for the abovementioned broad approach to valorization, this second group of indicators should also include collaborative R&D activity in progress, which is generally formalized in contract agreements. The taxonomy below summarizes these different elements, according to the level of maturity in the process of science and technology (S&T) – on the horizontal axis – and to the eventual existence of industrial partnerships possibly generating revenue – on the vertical axis (Table 7.1). It helps to specify the boundaries of valorization in a broad sense and through its various channels (quadrants ②, ③ and ④), as opposed to a more classic logic, where academic science largely ignores industry contacts ('Ivory tower' syndrome) or prefers to preserve its autonomy in the name of 'open science', i.e. when academic knowledge diffuses essentially via scientific publications or the mobility of researchers (quadrant ①). Of course, this classification is schematic and it probably neglects the existence of hybrid cases combining different logics, channels and indicators, at different points in time. For instance, a given academic invention can lead both to patent filings and to scientific publications; and, to go beyond a mere static approach, it should also be taken into account that a certain proportion of

with a large number of respondents (365 universities and 65 other PROs) from all of the 27 EU Member States and from 9 of 12 Associated States (European Commission, 2012). Last, similar surveys have been conducted for several years by the Australian government (NSRC: National Survey of Research Commercialization) and, in Japan, by the University Technology Transfer association (UNITT) and by the Ministry of Education, Science and Technology (MEXT, 2010).

Table 7.1 A taxonomy of the main stages, channels and indicators of valorization (or diffusion) of public research results, according to the level of maturity in the process of science and technology (S&T) and to the degree of proximity with industrial partners

		Level of maturity in the process of S&T	
		Relatively low level (S&T activity in progress)	Higher level (with already existing results)
Degree of proximity with industrial partners	**Low degree:** valorization or diffusion with no direct revenue generation and often without formal links with industrial partners	Pre-competitive 'Ivory tower' or 'open science' research (scientific publications, mobility of researchers) ①	Commercial potential (invention disclosures, patents filed or already granted) ②
	High degree: Valorization in the sense of finding industrial partners and generally of generating revenue	Collaborative R&D (contract agreements) ③	Actual use (spin-offs, licences) ④

Source: Author's taxonomy, building partly on Arundel and Bordoy (2008, p. 22).

invention disclosures (or patents) representing a commercial potential at a given point in time will in a later phase be transformed into actual use and revenue flows through the creation of spin-offs or licensing deals, while an existing start-up or an existing licensing deal can bring subsequent R&D agreements with the same university, etc. The arrows in Table 7.1 account for the possibility of such inter-actions or successive sequences.

7.3.1 Indicators of commercial potential (invention disclosures and patent numbers)

One of the main performance indicators related to commercial poten-tial is the number of invention disclosures, which reflects the volume of the research portfolio in which the respective PROs are involved, and which is often considered as the most important input for the con-cerned TTOs (Hülsbeck *et al.*, 2011). When judging by the average number of invention disclosures per TTO, Japan and the USA are well ahead of Europe, Australia and, to a lesser extent, Canada, according to the available surveys (Table 7.2). However, the reliability of this comparison is limited, notably by the fact that invention disclosure is mandatory for PROs in the USA but not in many other cases and notably in Europe (Conti and Gaule, 2011).

In relative terms, the gap between the USA, Canada and Europe is similar for the number of patents filed, so that, in the case of the concerned TTOs on both sides of the Atlantic Ocean, around every second invention disclosure is followed by a patent filing (Table 7.2). However, this indicator of patent filing also has its own drawbacks. One of the major limitations stems from the fact that, in many coun-tries, the PROs do not themselves own the patents resulting from collaborative R&D in which they are involved, considering that patent filing is expensive or that they often do not have the necessary enforce-ment capabilities, in case of litigation. This is why, in countries like Germany, France, Italy or Sweden and in contrast to the USA, a very large share of university-invented patents are filed under the name of the research partners – mostly firms – and thus are not owned by the universities concerned themselves, even though things have changed in this regard during the last decade (Lissoni *et al.*, 2008). Considering this phenomenon, Europe as a whole seems not to lag behind the USA in relative terms, for university patenting.

Table 7.2 Key performance indicators for university TTOs: an international comparison

	Europe				USA	Canada	Australia	Japan
	CEMI	ASTP	ProTon Europe FY	EKTIS	AUTM	AUTM	NSRC	UNITT/ MEXT
	FY 2007	FY 2008	2010	FY 2010	FY 2010	FY 2009	FY 2007	FY 2007
Number of survey respondents	211	99	295	365	183	≈30	77	74
Average staff (FTEs) per respondent	10.8	10.7	7.7	12	11 (in 2008)	9.9	6.4	15.8
Average number of invention disclosures per respondent	n.a.	36.7	19.3	29.9	112.8	51.9	16.3	128.4
Average number of priority patent applications per respondent	n.a.	13.8	9.3	14.4	67.1	23.6	10.9	84.5
Average number of spin-offs created per respondent	4.1	2.5	1.9	3.1	3.5	1.3	0.5	4.0
Average number of licences executed per respondent	7.8	13.0	16.0	14.5	29.3	16.8	7.6	18.2
Average licensing revenues (non-zero values; thousands of Euros)	n.a.	n.a.	458	715	n.a.	n.a.	n.a.	n.a.
Total licensing revenues (million Euros)	n.a.	89.2 (in 2007)	66.9	202.4	1,818.5	44.5	136.3	9.5

Note: For more details about these surveys and the data providers, see note 1 above.
Sources: Piccaluga *et al.* (2011, 2012), Balderi *et al.* (2010) and European Commission (2012).

7.3.2 Indicators of actual use (spin-offs, licensing activity) and collaborative R&D (contract agreements)

In terms of actual use, a key performance indicator is the creation of spin-offs, i.e. of start-ups emanating from universities or other PROs. Regarding the average number of spin-offs per TTO, the performances of Japan, Europe and the USA are globally very close to each other, according to most surveys, and Europe even seems to have the lead, when considering the results of the CEMI survey (Table 7.2). Admittedly, here again, a comparison based on this single criterion is possibly biased, because the policy-makers in the different countries do not usually place the same importance on the creation of spin-offs as a channel of knowledge or technology transfer. For instance, creating academic spin-offs has received higher priority in the UK than in France during the recent period (Mustar and Wright, 2010), and the TTOs in Italy generally consider the creation of spin-offs a more important goal than patenting or cooperating with industry (Algieri *et al.*, 2011), while the USA seem to focus more on licensing activity (Piccaluga *et al.*, 2011). Furthermore, the sheer number of spin-offs created says nothing about their viability (survival rate) and their importance in terms of jobs or value added (DeVol *et al.*, 2006).

Concerning licensing activity, interesting indications are firstly provided by the number of licences. It should be recalled that, given that patent filing and renewal fees are costly, patenting cannot be considered as a goal in itself, and that in many cases PROs patent their inventions only once they have identified potential licensees and after having negotiated a licence contract with an industry counterpart (Conti and Gaule, 2011). According to the available surveys, the average number of licences executed in 2009/2010 was per TTO globally twice higher in the USA than in Europe and in Canada (Table 7.2). In a similar perspective, it would be useful to compare also the number of R&D agreements per TTO, but the corresponding data are usually missing, with few exceptions like the recent EKTIS survey. Anyhow, all quantitative indicators mentioned so far – number of patents, licences, spin-offs, etc. – have a severe limit: they give useful indications about the volume of potential or actual transactions, but none on the economic *value* of the underlying knowledge transfer. Similarly, all the performance indicators considered up to this point are partial.

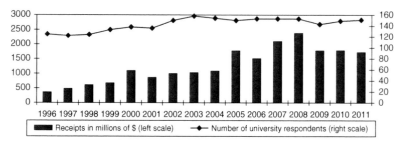

Figure 7.1 Licensing income in US universities (1996–2011). *Source:* Author's figure based on Roessner *et al.* (2009) and AUTM survey data.

7.3.3 Licensing revenue as an appropriate synthetic indicator of valorization performance?

What about a more synthetic criterion that could give an overall indication about the economic importance of the different aspects of valorization taken all together? A possible candidate would be the total receipts generated by the valorization activities of universities and other PROs. Available data on that aspect most often focus on licensing revenue, with a more or less broad definition.[2]

In the USA, as shown by the surveys regularly conducted by the AUTM, the total yearly value of the licensing receipts earned by the responding universities has grown from under $400 million in 1996 to almost $2.4 billion in 2008 and – probably due mainly to the effects of the global economic crisis – has since then diminished to under $1.8 billion in 2011. The fivefold increase observed over this total 16-year period is due only to a small extent to the fact that the number of respondents increased, especially since this number is roughly stable for more than a decade (Figure 7.1).

Even for the USA, however, it would be misleading to interpret such an increase as the reflection of a steady and widespread reality, as most of the gains in question are highly concentrated in the universities active in the biomedical field. Even for those universities, the amounts of these receipts are very irregular, and increases in recent years are

---

[2] In the case of the USA, the AUTM survey data include not only running royalties from the sale of licensed products and cashed-in equity from the sale of equity in the licensee received as part of the license consideration, but also all other types of gross license income such as upfront fees, annual minimum royalties, milestone payments, litigation settlements, etc.

often due to a 'jackpot/lottery effect' limited to a small number of beneficiaries and to exceptional transactions leading to a tenfold increase of usual revenue. As several cases observed in the recent period have shown, these 'big hits' correspond to one-time deals and huge lump-sum payments where the academic institutions concerned have either accelerated receipt of the future royalty streams through a sale of their royalty rights,[3] or have sold equity stakes, or have earned money from legal settlements (Abrams *et al.*, 2009). Anyhow, the distribution of licensing income in the USA is highly skewed. Only 10% of all university having responded to the AUTM concerning FY 2010 – that is 15 out of 149 – earned more than $38 million with licensing, and accounted together for 70% of the total licensing income of the whole sample. The ranking of the top universities occurs nonetheless to be relatively stable from year to year, at least over the period 2008–2010 (Figure 7.2), which testifies to the very strong capacity of these elite universities to commercialize their intellectual property.

7.3.4 Is the ratio of licensing revenue to research expenditure a more accurate productivity measure?

Of course, an international comparison in this domain must take account of the fact that the size of the different PROs can vary considerably between and within countries. Given that R&D expenditures is the most common measure of research input, at the level of countries or individual PROs, using the ratio of licensing revenue (royalty) to research expenditure seems to be an appropriate way to normalize the measure. This kind of ratio is also considered sometimes as a partial productivity measure to assess the valorization performance (Arundel and Bordoy, 2010; Heisey and Adelman, 2011).

For the fiscal year 2010, the evidence for the USA shows that, in percentage of research expenditure too, the distribution of licensing revenue shows a very large asymmetry (Figure 7.3). On top of the

[3] In the case of New York University, for instance, earnings in 2007 (nearly $800 million) primarily stem from the sale of the rights of the anti-inflammatory drug Remicade to the New York-based Royalty Pharma (worth $650 million). Annual revenues collected by this university are generally around $100 million in recent years. The situation is similar for Northwestern University, which earned more than $800 million in 2008, including $700 million through the sole sale of the rights of the 'blockbuster' anticonvulsant Lyrica to Royalty Pharma, while its usual annual revenue used to amount *at most* a few dozens of millions during the previous years.

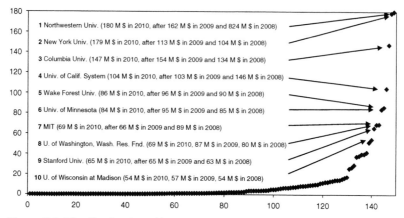

Figure 7.2 The distribution of licensing income among 149 US universities (in $ millions; FY 2010). The vertical axis of the chart shows the value of licensing income for each of the 149 universities listed on the horizontal axis in increasing order of licensing income. *Source:* Author's figure based on AUTM data.

distribution, 20 universities – among which some of the most famous and largest US universities (Columbia, Northwestern, Caltec, Stanford, University of Washington, etc.) – have reached very high levels, exceeding the threshold of 5%.[4] The ratio is slightly below that for the MIT, and many other prestigious universities like Harvard, Princeton, Yale, Cornell or Johns Hopkins have a much lower position in this ranking, with a ratio closer to 1% for most of them. Moreover, the median value is 0.9%, which means that the ratio in question is less than 1% for half of the sample.

In this domain, there is undeniably a high contrast between the USA and other comparable countries. The average ratio of licence revenue to research expenditures has roughly doubled in the US universities in recent years – from around 2% in 1996 to a 3% to 5% range during the second half of the last decade. Conversely, the corresponding value has gradually declined from around 2% to 1% in Canadian institutions (universities, hospitals and colleges) during the 2000s, while it lies between 1% and 2% for the higher institutions in the UK, has stagnated at around 1% in France for the group consisting of universities and the French National

---

[4] Previously, this was the case for 23 universities in 2009, 22 in 2008, 19 in 2007, 17 in 2006, 18 in 2005 and 16 in 2004.

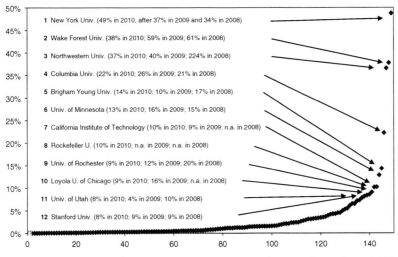

Figure 7.3 The ratio of licensing revenue to research expenditure in 149 US universities (in %; FY 2010). The 149 universities listed on the horizontal axis are sorted in ascending order according to the values of the ratio shown on the vertical axis. *Source:* Author's calculations based on AUTM data.

Centre for Scientific Research (CNRS),[5] and has fluctuated between 1.3% and 4.1% in Australian universities, publicly funded research agencies and medical research institutes (Figure 7.4). For Europe as a whole (including the UK) in 2008, the values published from the last ASTP survey (Arundel and Bordoy, 2010) correspond to a ratio of 1.24% for around 40 universities and to 1.46% for some 20 other PROs.

Of course, the comparability of these different data sets is limited by different factors.[6] It is nevertheless clear that the USA globally

---

[5] Due to the large number of mixed units consisting of researchers from university and from the CNRS, it would be less meaningful to present here a separate count. '90% of CNRS personnel are employed in laboratories located in the universities' (Mustar and Wright, 2010, p. 47).

[6] These surveys often differ according to the respective target populations (universities, other types of higher education institutions, hospitals or other types of PROs), the definition of licensing income and research expenditures, or the treatment of the missing values; see Arundel and Bordoy (2008). Another possible source of statistical artefact can be differences in the relative size and representativeness of the respective samples. As noticed in a study by the Milken Institute (DeVol *et al.*, 2006), the higher response rate in the AUTM surveys – as compared to most other surveys – can also induce an upward bias for the USA, given that for instance 96 of the top 100 US research universities participated in the survey for FY 2004.

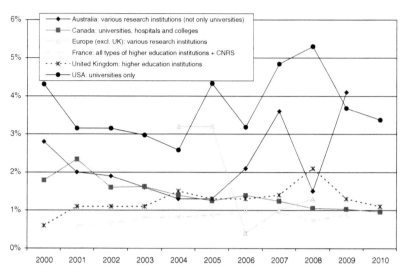

Figure 7.4 The ratio of licensing revenue to research expenditure in Australian, Canadian, European and US universities. *Sources:* Author's figure and calculations based on AUTM data (for the USA and Canada), on data from the French Ministry of Higher Education and Research (for France), on the HE-BCI (Higher Education – Business and Community Interaction) survey conducted by the HEFCE (Higher Education Funding Council for England) for the UK. The ratios concerning Australia (National Survey of Research Commercialization) and – on the basis of the ASTP survey – Europe (excluding the UK) stem from the Australian government (2011).

outperforms most other countries concerning licensing income, for a given level of research expenditure. But it is also apparent that for all countries considered, licensing revenue has often a high variability, no clear time trend and generally does *not* by a long way represent a substantial net contribution to the funding of the corresponding institutions.

Beyond licensing revenue, several studies globally comparing these countries also rely on standardized measures considering several output indicators (e.g. patent numbers, etc.) per million € (or $) of research expenditure or per 1,000 research staff. Based on the ASTP survey (DeVol *et al.*, 2006; Arundel and Bordoy, 2008, 2010) or on the EKTIS survey (European Commission, 2012) and referring to AUTM survey data, they show that, if European public research organizations are globally outperformed by their US counterparts for many of the usual indicators (but not for the number of licensing agreements), the hierarchy is reversed for the number of spin-offs, a dimension where Europe has the lead over the USA, Canada and Australia.

As for Conti and Gaule (2011), they go beyond simple descriptive statistical analysis and offer precise explanations. Their econometric study is based on both the CEMI survey and the AUTM data, and controls for factors such as size (i.e. academic science production measured by the number of scientific publications).[7] It shows that if, all other things being equal and compared to US universities, their European counterparts undeniably obtain lower licensing revenue, this is largely due to the way the TTOs are funded and organized in the USA, where the TTOs' staff has more experience in industry and has greater flexibility in managing its budget (Conti and Gaule, 2011).

All in all, the available evidence shows that, globally, PROs do not (so clearly) exploit their research results better in the USA than in other countries. It also leads to the point that, in order to assess the performance of valorization efforts in international comparison, it is necessary to understand how it is shaped by structural and institutional factors.

7.4 Performance gaps are largely explained by structural and institutional factors

Beyond more and less significant performance gaps in global quantitative terms, the international comparison reveals another striking fact, namely the large and persisting diversity of the respective national contexts, as well as of the practices and organizational structures of the corresponding TTOs.

7.4.1 Differences persist between institutional contexts, in spite of recent reforms

To explain international differences in the domain of public research valorization, a first major factor concerns the legal framework, which plays a major role for questions of governance and incentives. In this regard, the USA is an important reference, largely since the famous

---

[7] Using AUTM data and the unique CEMI survey conducted in 2008 concerning European universities (see note 1 above), Conti and Gaule (2011) claim that their study is the first to employ micro-level evidence to analyse transatlantic differences in this matter. Their regressions explain licensing performance (alternatively licence income and the number of licences) by controlling not only for factors concerning the level of resources available (such as the number of scientific publications, the size of the TTO staff, etc.) but also for the way the TTOs are managed (degree of orientation towards revenue generation, employment of staff with experience in the industry sector).

Bayh–Dole Act. This law, enacted in 1980, permits universities and research institutes to claim patent rights in inventions developed with federal funding, encouraging them therefore to establish or expand their respective TTOs. Before that, the situation was quite different and more complex, with a non-uniform system of rules concerning intellectual property ownership (Sweeney, 2012), so that US universities themselves were less often and less systematically involved with technology transfer through patenting and licensing.

Scholars often disagree about the specific purpose(s) and the different effects of the Bayh–Dole Act. However, the major purpose of the Bayh–Dole Act was clearly to promote the utilization of inventions, and *in fine* to produce the greatest public benefit for the taxpayer (Stercks, 2011), by integrating academic research into the USA's innovation ecosystem (Abrams *et al.*, 2009) and, hence, by contributing to an enhanced competitiveness. Indeed, Bayh–Dole requires that the federal funding recipients prefer domestic industry for the manufacture of their inventions (Sweeney, 2012).

As for the effects of the Bayh–Dole Act in the USA, they are obvious and impressive on the number of active TTOs (Stercks, 2011). But they are less clear on the patenting and licensing activities emanating from US universities, because if academic patenting and licensing undoubtedly increased a lot since the Bayh–Dole Act – notably in biomedical fields and some fields of engineering – the surge in this domain had already started before and can also be explained partly by other factors: the growth in federal financial support for basic biomedical research, the related rise of research in biotechnology, court rulings and shifts in federal policy that promoted an easier patenting in the biomedical field, etc. (Mowery *et al.* 2001). The fact is that, compared with other countries, it is easier in the USA to patent genetic material and, more broadly, research tools (DeVol *et al.*, 2006). Considering legal factors, another reason for the large number of academic patents in the USA is the fact that, independently of the Bayh–Dole Act, the patentability of inventions involving software is much less restricted there than in other countries and notably than in Europe (von Ledebur, 2006).

Anyhow, the Bayh–Dole Act has inspired a series of reforms in comparable countries, but almost 20 years later and starting from quite different initial situations. In the UK, most PROs are for more than two decades used to claim ownership of intellectual property (IP) created by their researchers (see Lambert Review, 2003) and the ownership rights question has been confirmed by the UK Patent Act of 1997 (Mustar and Wright, 2010). In Japan, a law enacted in 1998 aimed at promoting technological links between universities and private businesses, and induced the creation of many TTOs. In Italy, the national law establishing TTOs dates back to 1999, and the number of TTOs went up from five in 2000 to 58 in 2008 (Algieri *et al.*, 2011). In France, the main move in this direction was the law on innovation and research passed in 1999 ('loi Allègre'), which allowed academic researchers to transfer their results by creating start-ups, and which led in particular to the creation of the so-called 'industry and trade activity services' (SAIC: Services d'Activités Industrielles et Commerciales) in many universities, to facilitate the valorization and transfer of their research results. Moreover, the French universities have been granted organizational, financial and staffing autonomy only since a law passed in 2007 ('loi Pécresse') (Mustar and Wright, 2010).

In Germany too, several governments have successively striven to improve the organization of science–industry links, partly inspired by the US Bayh–Dole Act. A major dimension of this reform was in 2002 the repeal of the 'professors' privilege',[8] which shifted the ownership of academic research results from inventors to the university.[9] Since then, university researchers have been obliged to disclose their inventions to their respective university. One year before, in March 2001, the federal government had launched a 'valorization offensive' (*Verwertungsoffensive*). Similar to the French law of 1999, this German reform aimed at accelerating the commercialization of inventions derived from public research organizations and notably from universities, and has led to strengthening the role of patents in these institutions, which have been encouraged to develop their activity of the field of research commercialization.

In the respective countries, the considerations that motivated these reforms are probably at least as various as has been the case for the

---

[8] In Germany, this provision had long given university professors the right to decide by themselves if and how the results of their research should be exploited, so that many universities were themselves not much involved or supportive in this matter. Moreover, many of these professors did generally not care to patent their inventions and to manage these patents, or considered that they were unable to do so.

[9] 'Somewhat paradoxically, even though abolishing the professors' privilege was motivated by the apparent success of the Bayh–Dole Act, in effect the reform did not allocate the IPR in university inventions closer to the inventor, as Bayh–Dole Act had done, but rather removed them from the inventors to their employers' (von Proff *et al.*, 2012, p. 2).

Bayh–Dole Act in the USA. In the UK, for instance, policy-makers have put great emphasis on the creation of spin-offs as a channel of third-stream financing, with the underlying idea that allowing universities to be shareholders in such spin-offs would bring additional financial resources to the university; while the rationale is quite different in France, where the universities seldom take equity in these new firms, and where the development of academic spin-offs is more seen as part of a technological entrepreneurship policy (Mustar and Wright, 2010).

Anyhow, these institutional factors play an important role and explain to a large part why, from one country to another, the TTOs can differ in terms of practices and organizational structures.

7.4.2 Different missions and organizational structures for the corresponding TTOs

If the organizational structure of the TTOs should be adapted to the institutional context of the respective countries (Debackere, 2012), the appropriate choice is not always obvious, and several options can coexist within a given institutional arrangement.

This is for instance the case for the question whether the TTO should be internal or external to the public research organization concerned. In this respect, the most widespread model is the former, where the TTO is embedded within the PRO and dedicated to it, a choice largely motivated by the wish to promote a close proximity with the PRO's researchers (Debackere, 2012). This is the case for most TTOs in the USA (Abrams *et al.*, 2009). It is also the dominant model in Germany, but some scholars emphasize the flip side of the coin: the TTOs act then rather 'as a part of university bureaucracy than [as] proactive units in the technology transfer process' (Hülsbeck *et al.*, 2011, p. 13). The scene is different in other countries, notably in the UK and Australia, where TTOs are much more frequently organized as independent corporations operating in the name of the PRO, but sometimes functioning with a clear profit-oriented profile.[10] In the USA, an example of such an external institution is the Research Corporation, an organization which was founded around

---

[10] 'The extreme example of an independent corporation is Imperial Innovations, PLC, the technology transfer arm of Imperial College, London ... Imperial Innovations has a fiduciary responsibility to its shareholders to maximize its profits, and can no longer hew to the university's charitable mission' (Abrams *et al.*, 2009, p. 26).

1912 and used to supply patenting services to many universities (including the MIT) till the 1980s, but was focused on technology transfer itself rather than on profit.[11] Anyhow, such TTO structures which are not an integral part of the institution but independent corporations often lead to mixed results, according to Debackere (2012).

A related source of diversity stems from the respective status and mission of the TTOs and of the PROs concerned. More precisely, the global orientation of the valorization activities depends not only on the status (public vs. private) of the PRO but also on its type and its specific profile in terms of technological specialization. In France, for instance, the Institut Pasteur is a non-profit private foundation dedicated to public health worldwide and, as such, wishes to preserve its independence; this explains why almost a third of its budget stems from own resources generated by the valorization of its research activity (industrial royalties and receipts from research contracts). In the case of the USA, a survey conducted by Abrams *et al.* (2009) of some 130 US institutions shows that the reporting TTOs are more likely to be profitable in the case of private institutions – as compared with public institutions – and that research institutes' TTOs are more often profitable than those at universities and hospitals.[12] Similarly, an econometric study shows that the potential to generate returns from licensing activities is limited in the USA for public universities without a medical school (Bulut and Moschini, 2009).[13] In the USA, basically, only universities with schools of medicine or, to a lesser extent, of engineering or science are able to generate large amounts of invention disclosures (Jensen and Thursby, 2001). And if life sciences patents (especially in biotechnology and biopharmaceuticals) have an enormous potential value when they are licensed exclusively, the situation is very different in other domains like engineering, where the patent is often not considered as an essential vector of technology transfer (Kesan, 2009).

[11] See Kesan (2009), as well as the website of the Research Corporation for Science Advancement (www.rescorp.org/about-rcsa/history).

[12] It is also interesting to note that a very large proportion (41.4%) of the responding TTOs have no formal mission statement (Abrams *et al.*, 2009).

[13] As a private university, Wake Forest University (North Carolina) provides a good example of this particular logic: the mission of its TTO (Office of Technology Asset Management: OTAM) is explicitly 'to maximize the value of Wake Forest University's and Wake Forest School of Medicine's intellectual assets through the creation of novel and effective models for commercializing technology' (www.wakehealth.edu/OTAM/).

7.4.3 A large variety of practices in terms of licensing and spin-offs

The degree of variety is also high concerning the actual practices of the TTOs. One reason for this is that policy-makers and scholars often have contrasting views about the best way to valorize public research results. On the one hand, patenting and licensing has the preference of many TTOs, especially in biomedical fields (Stercks, 2011) and more broadly in the case of early-stage technologies (Heisey and Adelman, 2011), when there is very great difficulty in accessing *ex ante* the quality and the economic potential of the underlying inventions. On the other hand, spin-offs are in many other cases privileged as an entrepreneurial way to commercialize academic research, leading to renewed industrial structure (Algieri *et al.*, 2011), notably by enhancing competition in existing industries (Debackere, 2012). Incentives also play a role in the choice of spin-offs vs. patenting and licensing, because the licensing strategy is often preferred when academic researchers get themselves a substantial share of the licensing receipts potentially generated by their patented invention; conversely, spin-offs occur more frequently when the PRO concerned prefers to receive compensation for the efforts of its researchers in the form of an equity interest in the start-up firm (Kesan, 2009). But licensing remains often considered as the quickest way of transferring technological knowledge from academia to the business sector, with a higher probability of getting technology to market and requiring less resource, as compared with promoting spin-offs, which involves funding and building businesses from scratch, as shown in a report on business–university collaboration commissioned by the UK Government ten years ago (Lambert Review, 2003).

A similar dilemma can be found concerning the choice between exclusive or non-exclusive licences, as both of them have pros and cons. All other things being equal, exclusive licences are usually preferred, when the TTO can identify a single, credible potential licensee interested in the technology concerned, while the TTO is conversely inclined to grant non-exclusive licences in other cases, in order to promote competition between the different licensees (Abrams *et al.*, 2009). It depends to a large extent on the technological domain considered. For instance, the balance tends to tip in favour of non-exclusive licences in the case of embryonic technologies and of enabling technologies – that is of upstream inventions – in order not to impede further downstream developments. In other cases, exclusive licences can be preferred when very important investments are necessary to develop the technology – as is the case in the pharmaceutical industry (Kesan, 2009) – or when the TTO wishes to maximize its royalties (Stercks, 2011). However, TTOs can sometimes earn more in royalties with non-exclusive licences, as shown by the famous case of the Cohen-Boyer patent on recombinant DNA technology, for which Stanford University deliberately chose not to charge a high level of licensing fees and royalty rates (Kesan, 2009; Stercks, 2011). As a matter of fact, the AUTM data show that exclusive licences are no more preponderant on the part of TTOs in the USA: the share of non-exclusive licences has been on average around 60% since the beginning of the previous decade. But exclusive licences are slightly more often applied than non-exclusive licences among European TTOs, according to the EKTIS survey (European Commission, 2012).

Despite similarities concerning recent changes observed on such aspects, the landscape in the domain of valorization of public research remains marked by a great diversity worldwide. A 'one-size-fits-all' approach would therefore not be appropriate, in two respects. Firstly, an international convergence in public policies is unlikely, because differences across countries matter and persist, due to path dependence stemming from national systems of innovation, as suggested by Mustar and Wright (2010) on the basis of a comparison between the UK and France. Secondly, differences also remain between the different PROs within countries, because the practices of the various TTOs have to be attuned to the missions of their respective PROs, as well as to their scientific and technological profile (Kesan, 2009). This diversity is for instance very clear in the case of the Japanese university technology licensing offices (Senoo *et al.*, 2009). Yet, the different TTOs also face common challenges, so that lessons can be learned from their diverse practices.

7.5 Common challenges beyond the difficulty of comparing and assessing performance

Even though TTOs are generally embedded in quite specific contexts concerning the academic disciplines of their respective PROs or the geographical scope of the industrial partners they deal with, they have to overcome similar obstacles in order to meet success, and the ultimate goals they pursue must be clarified.

7.5.1 Which factors are critical for the success of TTOs?

The success of TTOs usually depends on a series of critical factors. The most important is arguably the involvement of the concerned academic researchers (von Ledebur, 2006). The main reason for it is that TTOs mostly deal with early-stage inventions. Hence, even after a licensing agreement has been signed with an industrial partner, the underlying technology needs to mature and therefore will only acquire a substantial economic value after a subsequent phase of development in cooperation with the concerned researchers, i.e. with the persons who usually know best the full potential of their inventions (Jensen and Thursby, 2001). It implies that the TTO staff must operate in close proximity to these academic scientists (Debackere, 2012). Now, this cannot be taken for granted, given that, in countries like Germany, university researchers are frequently sceptical or reluctant vis-à-vis the TTOs (Hülsbeck et al., 2011) or vis-à-vis the valorization agencies (Fritsch et al., 2007). The fact is that the dominant norm of academia is traditionally 'open science' – with a preference for scientific publications as a way to widely disseminate research results – and that academic inventors are usually driven to a large extent by other motives than financial rewards (Kesan, 2009). Changing this situation would require not only the recognition of the valorization dimension as an important aspect for the career of the researchers (von Ledebur, 2006) but also that the researchers have enough incentives to cooperate in the phase following the licensing contract (Debackere, 2012).

Of course, another crucial factor of success is the capacity of TTOs to cooperate closely with the private sector. It goes without saying, but it is difficult to implement, because it requires having a highly qualified and deeply specialized personnel, with a special blend of skills in the domain of technology, management and law, notably concerning IPR. Moreover, the TTO staff must both develop a multidisciplinary approach and have sectoral competences, with a fine knowledge of the potential licensees in the industries concerned (Debackere, 2012). A personnel with such a professional culture remains a rare resource in countries like Germany, where – with a few exceptions like Ludwig Maximilian University (LMU) Munich and Technical University (TU) Dresden – most universities lack the 'entrepreneurial spirit' that is necessary to proactively foster technology transfer, according to Hülsbeck et al. (2011).

The importance of this professional culture explains why the past experience of the staff plays a major role in the performance of TTOs.

In the case of the USA, the evidence generally confirms this learning-by-doing effect, showing that there is a positive correlation between the profitability of the TTOs and their age. In this regard, the fact that TTOs have in average been established for a longer time in the USA than in most other countries could be a reason why the US TTOs are usually more profitable.[14] Of similar importance is the issue of scale, because the performance of TTOs is also positively and significantly correlated with the number of their personnel. It is obvious that small TTOs in most cases lack the resources needed to achieve a certain level of performance, notably in terms of profitability (licensing revenue). Furthermore, this problem of critical mass arises concerning not only the (human or financial) resources of the TTOs themselves but also the scientific resources of the corresponding PRO, measured for instance by its research budget or by the number of its invention disclosures.[15]

In the case of the TTOs in the USA, the evidence shows that 'increasing TTO size [is] a substitute for early entry into self-managed university technology transfer' (Heisey and Adelman, 2011, p. 54). This is probably why policy-makers in other countries sometimes wish to buy time by addressing the issue of size. During the last 12 years, this has been the case in Germany and France, with the attempt to mutualize resources by creating shared valorization services at a regional level (Box 7.1). In the UK, the Lambert Review of Business–University Collaboration published in December 2003 made a similar recommendation (Lambert, 2003). Another slightly different – and probably less costly – solution to this problem of critical mass would require the networking (or clustering) of the existing TTOs on a regional scale, through a 'virtual, collaborative model of TTO organization', as suggested by Debackere (2012).

But does it make sense to try to speed up the improvement of valorization services by investing a lot in financial and human resources within a short period of time? There are grounds for being sceptical about this. For, according to several studies, the relationship between TTO size and TTO output is subject to diminishing returns to scale (see

---

[14] In the sample of US universities studied by Heisey and Adelman (2011), 15% report that they already had a TTO before the Bayh–Dole Act (1980). In European countries, most universities have far less experience working in close relationship with the economic world. Conti and Gaule (2011) confirm that the age of university TTOs is on average significantly higher in the USA than in Europe.

[15] Concerning the role of size and age for the performance of the TTOs in the USA, see Heisey and Adelman (2011), Abrams *et al.* (2009) and Kesan (2009), as well as previous studies mentioned by von Ledebur (2006).

Box 7.1   The creation of shared valorization services for a group of PROs: a major feature of a recent reform in Germany

In Germany – as in France a few years later – the state has realized that, in order to strengthen the capacity of universities and other PROs to enhance their performance, it is necessary to achieve a certain 'critical mass' in terms of skills. This is the reason why, in the framework of the valorization campaign (*Verwertungsoffensive*) launched in spring 2001, the federal government has promoted the creation of 22 patent and valorization agencies (*Patent- und Verwertungsagenturen*: PVA) throughout Germany, with usually one agency in each *Bundesland*. As external and largely autonomous service providers, these agencies are usually responsible for a network of (mostly academical) organizations within a region. Accordingly, they adapt their specialization domain to the profile of the organizations for which they are active. They also support the TTOs which pre-existed at the level of the various PROs, but were – and still remain – often undersized and therefore unable to perform effectively. If these agencies most often have had de facto a strong regional dimension up till now, it is recommended that in the future they should specialize more in technology domains and have a stronger international orientation, particularly at the European level, which requires at least an integration into supraregional cooperation and networks. The federal government declares that these agencies should be self-financing in the long run but has acknowledged that the key issue is ultimately their macroeconomic utility through job creation, contribution to fiscal receipts or to subsequent innovation.

The first attempts to assess the performance of these German regional agencies are confounded by the fact that, at this level, the number of invention disclosures and patents filed rose sharply during the three years that followed the abolition of the 'professors' privilege' (i.e. during the years 2002–2004), but has stagnated since then. They generally conclude that the agencies have not been able to induce clear positive effects on the valorization activities of German universities yet, and that the relative importance of the transfer of technological knowledge from PROs to private firms remains globally insufficient.[16] Obviously, both the universities and the firms are most often unsatisfied with the activity of these

---

[16] In the German landscape, two exceptions are the valorization agencies of the Fraunhofer Society (applied research) and of the Max-Planck Society (basic research), which, in contrast to the other agencies, had already existed for more than 25 years before the recent reforms, and which have a central nationwide organization; see Fritsch *et al.* (2007) and von Proff *et al.* (2012).

Box 7.1 *(cont.)*

agencies, and at least a decade seems necessary to harvest the first substantial positive results.

These lessons deserve attention in France, where the government decided in 2010 to launch similar regional agencies (*Sociétés d'accélération du transfert de technologie*: SATT). In mid 2012, five had already been established and four others were in preparation. Like their German counterparts created a decade ago, these emerging tech-transfer companies are established as external service providers for a given territory, and they are supposed to be self-financing in the long term and to contribute to the funding of public research via the revenue generated.

Main sources: EFI (2012), von Proff *et al.* (2012), Fritsch *et al.* (2007), BMWi (2007) and von Ledebur (2006).

Conti and Gaule, 2011), i.e. exhibits an elasticity smaller than one. In the case of TTOs in the USA, for instance, an additional increase in the number of invention disclosures leads to a proportionally smaller increase in the number of licences, which itself generates a smaller increase in royalties at the margin (Thursby *et al.*, 2001). Similarly, Lach and Schankerman (2008) found that a 10% increase in the number of TTO professionals at US private universities raises licence income by less than 6%, and has no significant effect in US public universities.

In this respect, policy-makers should keep in mind the fact that valorization of public research results is a process that is usually combined with a rather long time horizon, particularly when basic research and early-stage technologies are involved, which is generally the case. This is confirmed by the experience of TTOs in the US universities. According to Abrams *et al.* (2009, p. 27), it took more than 13 years to establish the first wave of TTOs that were created after the Bayh–Dole Act, and then several more years were necessary not only to recruit and train the highly skilled employees needed but also to develop their professional culture; and a portfolio of inventions cannot be licensed overnight: 'it generally takes from one to four years to find a licensee to make a commitment to develop a technology'. Kesan (2009, p. 2180) states that, 'typically, it takes between five and ten years for a TTO to break even, if it does at all'. A similar lesson can be learned from the British and French experience. As explained by Mustar and Wright (2010), public authorities tend to underestimate the difficulties and the timescales in this domain, as well as the length of the learning process involved in academic organizations, where attitudes and culture can only be changed year on year.

7.5.2 The need to clarify the ultimate goals of valorization

Most difficult to predict and to manage, at the level of individual TTOs, is notably licensing revenue, which generally remains uncertain and fluctuating, given that only a small share of total patents and licences generate substantial income. In Germany, for instance, only one in every ten patents filed by universities is commercialized, and only one in every hundred patents results in significant licences (Fritsch *et al.*, 2007). Basically, this stems from that fact that a very large share of licensed technologies emanating from PROs are no more than a proof of concept: 48% in the case of the 62 US universities studied by Jensen and Thursby (2001).

In this context, is revenue maximization actually – and should it be – a major objective or even the ultimate goal, for valorization activities? In the case of the USA, this question is highly controversial.[17] The US TTO managers surveyed by Jensen and Thursby (2001) consider licence revenue as a slightly more important indicator of success than any other outcome (inventions commercialized, licences granted, sponsored research, patents), and they think that the university administrators – their paymasters – more or less share the same view. In the perception of these TTO managers, however, the faculty has a different scale of preferences, ranking sponsored research clearly in the first position, ahead of licence revenue. This means that TTO managers are generally aware that they have not only to deal with several objectives but also to balance faculty and administration objectives. The survey conducted by Abrams *et al.* (2009) among TTO managers in the USA (with 112 complete responses) tells a slightly different but not contradictory story, showing that the main driver of their office is more often 'providing a service to the faculty' (39.2%) and 'translating research results' (34.6%) than 'revenue maximization' (11.5%) or other considerations (14.6%). Abrams *et al.* (2009, p. 25) conclude that 'financial return is not the main factor in technology transfer organization and behaviour', and that 'only 10% of technology transfer activity in the US is driven by revenue maximization'.[18] Other scholars like Kesan (2009) are more critical and consider that technology transfer activities in the USA are

---

[17] For a review of literature, see Kesan (2009) and Stercks (2011).

[18] In this sample, moreover, only 2.5% of cases where there is a formal mission statement correspond to institutions that mentioned maximizing revenue or income (Abrams *et al.*, 2009).

Box 7.2 Which profitability for which proportion of TTOs?

There is no clear-cut notion of financial performance or profitability, in the domain of public research valorization, because an institution's TTO often has to share licensing income not only with the corresponding inventors but also with the inventors' laboratories, departments or colleges. Moreover, the activity of the TTOs involves not only potential receipts but also large amounts of expenses: patent filing fees, renewal fees to maintain granted patents in force, personnel costs, etc. As shown in the survey of Abrams *et al.* (2009) among 130 institutions (mostly universities but also a few federal laboratories), the reporting TTOs have contrasted positions in terms of financial performance or profitability: every second (52.3%) TTO is loss-making, one out of five (20.8%) is gross profitable (i.e. total income exceeds total expenses), 10.8% are net profitable (i.e. total income less distribution to inventors exceeds total expenses) and only 16.2% are self-sustaining (i.e. total income less distribution to inventors, colleges/laboratories, provost, university, etc. exceeds total expenses). In other words, TTO activity is barely profitable in gross terms or strongly unprofitable for the vast majority of PROs, and not only for the young ones; this is what several other studies have confirmed not only for the USA but also for the UK and Germany.[19] Far from being normal cases that should be considered as a benchmark,[20] the few universities with the top performance in terms of licensing income are in fact abnormalities, as stressed by Kesan (2009).

---

predominantly and too narrowly 'patent-centric and revenue-driven with a single-minded focus on generating licensing income and obtaining reimbursement for legal expenses' (Kesan, 2009, p. 2169).

Anyway, it would be illusory to think that TTO activity might or should generally be a net source of profit for the PROs concerned, since licence revenues are usually partly attributed to the inventors themselves and are also used to cover the costs of the TTO (Box 7.2).

---

[19] For more details or a review, see Stercks (2011), Heisey and Adelman (2011), Mustar and Wright (2009), Fritsch *et al.* (2007) and von Ledebur (2006). In the case of the German patent valorization agencies (PVAs) created a decade ago, total receipts have increased strongly since 2002 but reached only €4.9 million in 2010. These PVAs remain clearly in deficit and will also continue to rely on a core funding from the state in the future; during the period 2011–2013, they are granted around €8.5 million each year from the federal state and the *Länder* (EFI, 2012).

[20] 'A mean figure of licensing revenue as 3% of research expenditures, based on recent performance by some universities, is not a useful benchmark. Better

The fact that technology transfer activity is in most cases unprofitable *in financial and microeconomic terms* – i.e. is a net source of costs at the scale of the TTOs concerned – should therefore not necessarily be considered as a failure or as a sign of inefficiency.[21] Indeed, PROs have to bear the burden (costs) of knowledge transfer, while the revenue and welfare gains generated are mostly captured by private businesses and by society at large (customers, etc.), through various externalities. Policy-makers and academic administrators should accept this dilemma, take into account the positive *macroeconomic* impact of a well-designed valorization policy and, as explained by von Ledebur (2006), should provide the TTOs with sufficient funding.

Of course, 'indicators of TTO performance should not be confused with indicators of the societal impact of research' (Heisey and Adelman, 2011, p. 57), all the more as all PROs and hence all TTOs do not pursue identical economic objectives, as already shown: some of them are primarily eager to maximize the revenue gained from their research activity, while others are more oriented toward different aims. And licensing revenue is undeniably to be welcomed in the self-interest of the PROs concerned, if not in the general interest, insofar as the money is subsequently reinvested into research. In this perspective, additional licensing receipts are sometimes considered as a necessary contribution to the financing of research, in a context of limited public funds or of declining government funding (Algieri *et al.*, 2011; Stercks, 2011; Debackere, 2012).

Kesan (2009) is nonetheless right to state that increasing licensing revenue of the US universities has probably been a result of the Bayh–Dole Act but cannot be considered as its direct purpose. It is true that, beyond the special case of the USA, the rationale for public research valorization is ultimately to increase wealth and job creation, by encouraging 'the transfer of science and technology innovation to the business sector and society more generally' (Mustar and Wright, 2010, p. 45). In that sense, technology transfer is now depicted as the 'third mission' of the research universities – alongside education and research – and TTOs have been at the core of this new mission since the middle of

---

estimates for recent years would be about 1% for universities with purportedly efficient TTOs, or 0.6% for all universities' (Heisey and Adelman, 2011, p. 57).
[21] This idea can be challenged. According to Sweeney (2012), the main reason why TTOs are unprofitable is probably that they are *over*-patenting, investing too much time and resources on patents that do not deserve it.

the 1990s (Debackere, 2012). In other words, the main rationale of valorization activity is to promote utilization of results stemming from publicly funded research, not to maximize any financial return. Moreover, it seems that the wish to maximize financial return, which seems often illusory but obviously remains a temptation for several institutions in countries like the USA or the UK, would probably be in conflict with the broader goal of overall social–economic benefits.[22]

7.5.3 The need to acknowledge and limit possible negative impacts

A valorization policy with too much focus on patenting and licensing revenue might have serious drawbacks, insofar as excessive commercialism is harmful to academic research.[23] Excessive increased transaction costs and access charges could adversely affect research programmes. In this respect, a particularly controversial problem, notably in the USA, is the growing number of 'upstream' patents, i.e. the increased patenting of 'research tools', which serve as inputs to science. Privatization of these research results through a proliferation of upstream patents would lead to the so-called 'tragedy of the anti-commons' (Heller and Eisenberg, 1998), that is to a situation where an excessive number of rights in a resource could lead to underuse of this property, due to patent hold-ups, royalty stacking and the exclusion of players from certain research fields. Some scholars view licensing by academic organizations as amounting to double taxation, given than it stems from publicly funded research; according to Kesan (2009), these PROs would therefore do better to grant royalty-free licenses.

Another frequent concern is that the increasing importance of commercialization for the PROs could also distort the direction of academic research. Yet, the evidence is mixed in this respect, as shown by Larsen (2011), Stercks (2011) and Kesan (2009). On the one hand,

---

[22] 'Many UK universities are still seeking large financial returns, which is unrealistic and is likely to reduce the broader benefits of their research' (Lambert, 2003, p. 49). The title of a book recently published in the UK reflects this frequent belief in tech-transfer activities as a virtual cash machine: Richards G. (2012) *University Intellectual Property: A Source of Finance and Impact* (Petersfield, UK: Harriman House).

[23] For a discussion and a review of literature on this topic, see Sweeney (2012), Larsen (2011), Stercks (2011), Heisey and Adelman (2011) or Kesan (2009).

empirical studies do not support globally the assumption that the rise of 'academic enterprise' would induce a preference for applied research at the expense of basic research. On the other hand, there are serious grounds to think that the growing trends toward monetization of public research encourage the development of research with commercial potential, at the expense of more disinterested research or of inventions that cannot be patented. Similarly, such incentives and behaviours seem to affect negatively the sharing and diffusion of research results among academics. On the part of academic organizations, for instance, more patenting and more scientific partnerships with industry sometimes lead to delaying or avoiding the publication of scientific findings.

There is also – at least in the USA – a risk that universities, because of their closer relations with private businesses, tend themselves to act increasingly as businesses do, notably by enforcing their patents more frequently, through litigation, in a context where the common-law research exemption from patent infringement (experimental use exception) is not clearly recognized in courts. Moreover, and this is probably the case in many other countries as well, PROs are probably too often inclined to claim ownership on inventions that result from research partnership with firms (Sweeney, 2012).

All in all, there is a clear risk that excessive commercialism could undermine traditional academic values and be detrimental to the culture of 'open science'. In other words, there is potential conflict of interest between the logic of 'dissemination of new research findings' and 'the commercial appropriation of new knowledge' (Debackere, 2012, p. 6). That said, this conflict of interest is not fatal. The evidence shows that there is in most cases a complementarity between academic research and technology transfer, as exemplified by a positive relationship between scientific publications, on the one hand, and patenting and commercialization activities, on the other hand (Larsen, 2011). '[R]ather than being opposing forces, academic research and involvement in industry–science links tend to reinforce the core missions of both' (Debackere, 2012, p. 10).

Another way to avoid these potential problems would be to explore alternative or more comprehensive approaches to valorization. The basic underlying idea is that it is in many cases possible and preferable to promote the utilization of technological knowledge with less emphasis on patenting and licensing, as compared with the dominant practices (Box 7.3).

Box 7.3 Alternative or broader approaches to tech-transfer and valorization in the USA

In the case of the US universities, the experience of successful TTOs shows that it is possible to earn substantial revenue from valorization activity through various channels and without investing too much in patenting and licensing. According to Stercks (2011), a good example is Stanford, where there exists a simple and standardized procedure for material transfer agreements (MTAs) instead of systematic licensing contracts, and where academic researchers can choose to place their inventions into the public domain, when they consider that it brings a more effective commercialization. In a similar vein, Sweeney (2012, p. 311) recommends that 'underutilised patents should be placed in the public domain to be monitored and exchanged through a national tech-transfer office'. In other cases, other rules suggested by Stercks (2011) require changes either at the level of the PROs themselves, or in national law (for instance concerning the research exemption in the USA) or through an international agreement. As for Kesan (2009), he advocates for open collaborations, free participant use agreements, and royalty-free licensing,[24] adding that several additional and informal technology transfer methods have been identified by other scholars or professionals, such as 'working directly with industry personnel to commercialize technology, coauthoring publications with industry personnel, and serving as paid consultants to private firms' (Kesan, 2009 p. 2198). These alternative channels are based on the implicit assumption that academic scientists are aware of the need to valorize theirs results and are to some extent able to do it without intermediaries such as TTOs.

7.6 Conclusion

Regarding public research results, the core idea about valorization is legitimate. It reflects a sound concern about the utilization of inventions arising from publicly funded research by private businesses. But there is often a first misunderstanding about the notion itself.

---

[24] Royalty-free licensing corresponds for instance to the case where an academic spin-off has been created, and where the PRO obtains a share of this firm as a counterpart of the licence (Kesan, 2009). Therefore, this valorization channel does not necessarily underestimate the necessity of incentives to involve the academic researchers in the development phase following the licensing contract (as stressed by Jensen and Thursby, 2001).

A frequent conception of valorization is focused too heavily on the downstream dimension, namely the commercialization of IPR notably through patenting and licensing. Though, it is more realistic to adopt a broader approach which, as reflected by the actual practices of most TTOs, encompasses the upstream dimension of collaborative research, and corresponds to a much greater variety of tasks, ranging from contract agreements to spin-offs creation and coaching.

Concerning the various activities of these TTOs, the available data tend to be misleading at first glance, suggesting that the USA outperforms other industrialized countries for almost all criteria. But these results have to be interpreted with caution. When several possible sources of bias are taken into account, the international comparison leads to more mixed results. All in all, it shows that the only indicator for which the USA has a clear leadership in relative terms is the value of licensing revenue. Moreover, structural and institutional factors explain a large part of the performance gaps. These qualitative differences also imply that public policies concerning technology transfer and commercialization cannot follow a general pattern and must reflect the diversity of missions assigned to different research organizations in question.

Yet, a few general lessons can be learned from the experience of TTOs in diverse countries, as well as from the economic analysis. One of them is that patents and licensing play a crucial role as incentives in this matter, notably to promote the involvement of academic researchers in close and often long-lasting science–industry partnerships. But, in the domain of valorization as in many others, patents are of varying importance depending on the technological domain considered: they are generally an essential vector of technology transfer in biotechnology but not so much in engineering, for instance. Furthermore, the evidence shows that licensing income varies a lot in time and space according to many factors like chance, technology specialization or status of the respective research organizations, etc. Apart from a few large research organizations with particular strengths in life sciences and with well-staffed and trained TTOs, the vast majority of cases correspond to unprofitable valorization activities, at the level of individual TTOs.

And this is probably another general rule in the matter of valorization: size (scientific and human resources) and experience play a major role in explaining a high level of performance, beyond the sole issue of licensing income. But they substitute with each other only to a certain extent: money cannot buy everything and a long time horizon is usually needed to build the necessary know-how, as well as to develop extensive and deep relations with industry. Indeed, it would be peculiar to consider the development of strong relations between science and industry as a way to monetize public research results and possibly to generate financial income. The right approach is rather the other way around: valorization aims at increasing the utilization of these results through various channels involving firms. In this process, intellectual property and money are very important ingredients, but neither of the two should be an end in itself.

However, this is probably the second major misunderstanding about valorization: many policy-makers and scholars conceive of it as a source of revenue and therefore as a way to finance academic research. But actually, commercialization activities do not in general represent a systematic vector of funding. They are in most cases a cost factor, rather than a net source of income. At best, the pursuit of financial reward can make sense at the microeconomic level, when the organization concerned has a profit-oriented mission. But this is absurd in terms of public welfare, all the more so as an excessive commercialism tends to impede public research by undermining the ethics of 'open science'. Where necessary, the need to limit some of these possible negative impacts justifies exploring alternative approaches to practices focusing on systematic patenting, high royalty rates and exclusive licenses. For governments, the true rationale of valorization activity is to promote a wide utilization of results stemming from publicly funded research, not to maximize any financial return. This also implies that objectives of self-financing are illusory for patent and valorization agencies recently created at the regional level, in countries such as Germany and France.

References

Abrams I., Leung G. and Stevens A. (2009) How are U.S. technology transfer offices tasked and motivated: is it all about the money? *Research Management Review*, 17, 1–34.

Algieri B., Aquino A. and Succurro M. (2011) Technology transfer offices and academic spin-off creation: the case of Italy. *Journal of Technology Transfer*, 36, published online.

Arundel A. and Bordoy C. (2008) *Developing Internationally Comparable Indicators for the Commercialization of Publicly-Funded Research*, Working Paper 075. Maastricht, Netherlands: UNU-MERIT.

(2010) *ASTP Summary Respondent Report: ASTP Survey for Fiscal Year 2008*, Report produced by UNU-MERIT for the Association of European Science and Technology Transfer Professionals. Maastricht: UNU-MERIT.

Australian Government (2011) *National Survey of Research Commercialisation: 2008 and 2009*. Canberra: Department of Education, Science and Training

BMWi (Federal Ministry of Economics and Technology) (2007) Die Verwertungsoffensive: Ein wichtiger Baustein der Innovationspolitik. *Schlaglichter der Wirtschaftspolitik, Monatsbericht*, No. 09/2007, pp. 16–20.

Bulut H. and Moschini G. (2009) US universities' net returns from patenting and licensing: a quantile regression analysis. *Economics of Innovation and New Technology*, 18, 123–137.

Conti A. and Gaule P. (2011) Is the US outperforming Europe in university technology licensing? A new perspective on the European paradox. *Research Policy*, 40, 123–135.

Debackere, K. (2012) *The TTO, a University Engine Transforming Science into Innovation*, Advice Paper 10. Leuven, Belgium: League of European Research Universities (LERU).

DeVol R., Bedroussian A., Babayan A., Frye M., Murphy D., Philipson T. J., Wallace L., Wong P. and Yeo B. (2006) *Mind to Market: A Global Analysis of University Biotechnology Transfer and Commercialization*. Santa Monica, CA: Milken Institute.

EFI (2012) *Gutachten zu Forschung, Innovation und technologischer Leistungsfähigkeit Deutschlands*. Berlin: Expertenkommission Forschung und Innovation.

European Commission (2012) *Interim Findings 2011 of the Knowledge Transfer Study 2010–2012*, report prepared by empirica GmbH. Bonn/Maastricht/Solothurn: UNU-MERIT and FHNW.

Fritsch M., Henning T., Slavtchev V. and Steigenberger N. (2007) *Hochschulen, Innovation, Region: Wissenstransfer im räumlichen Kontext*. Berlin: Sigma.

Heisey P. and Adelman S. (2011) Research expenditures, technology transfer activity, and university licensing revenue. *Journal of Technology Transfer*, 36, 38–60.

Heller M. and Eisenberg R. (1998) Can patents deter innovation? The anticommons in biomedical research. *Science*, 280, 698–701.

Hülsbeck M., Lehmann E. and Starnecker A. (2011) Performance of technology transfer offices in Germany. *Journal of Technology Transfer*, 36, published online.

Jensen R. and Thursby M. (2001) Proofs and prototypes for sale: the licensing of university inventions. *American Economic Review*, 91, 240–259.

Kesan J. (2009) Transferring innovation. *Fordham Law Review*, 77, 2169–2223.

Kline S. and Rosenberg N. (1986) An overview of innovation. In R. Landau and N. Rosenberg (eds.), *The Positive Sum Strategy: Harnessing Technology for Economic Growth*, pp. 275–305. Washington, DC: National Academies Press.

Lach S. and Schankerman M. (2008) Incentives and invention in universities. *RAND Journal of Economics*, 39, 403–433.

Lambert, R. (2003) *Lambert Review of Business–University Collaboration*, Final Report. London: HM Treasury.

Larsen M. T. (2011) The implications of academic enterprise for public science: an overview of the empirical evidence. *Research Policy*, 40, 6–19.

Ledebur, S. von (2006) Patentverwertungsagenturen und der Wissentransfer von Hochschulen: ein Litteraturüberblick. *Wirtschaft im Wandel*, 12, 266–274.

Lissoni F., Llerena P., McKelvey M. and Sanditov B. (2008) Academic patenting in Europe: new evidence from the KEINS database. *Research Evaluation*, 16, 87–102.

MEXT (Japanese Ministry of Education, Culture, Sports, Science and Technology) (2010) *State of University Technology Transfer in Japan: FY2009*. Nagoya, Japan: Nagoya University, Inc.

Mowery D., Nelson R., Sampat B. and Ziedonis A. (2001) The growth of patenting and licensing by U.S. universities: an assessment of the effects of the Bayh–Dole Act of 1980. *Research Policy*, 30, 99–119.

Mustar P. and Wright M. (2010) Convergence or path dependency in policies to foster the creation of university spin-off firms? A comparison of France and the United Kingdom. *Journal of Technology Transfer*, 35, 42–65.

Piccaluga A., Daniele C. and Patrono A. (2012) *The ProTon Europe Annual Survey Report (Fiscal Year 2010)*, presentation for the conference on *Sustainability: Innovation Services for a Smarter Economy*, 25–27 April 2012, Copenhagen, Denmark.

Piccaluga A., Balderi C. and Patrono A. (2011) *The Proton Europe Seventh Annual Survey Report (Fiscal Year 2009)*. Pisa, Italy: Institute of Management, Senola Superiore Sant' Anna.

Proff S. von, Buenstorf G. and Hummel M. (2012) University patenting in Germany before and after 2002: what role did the professors' privilege play? *Industry and Innovation*, 19, 23–44.

Roessner D., Bond J., Okubo S. and Planting M. (2009) *The Economic Impact of Licensed Commercialized Inventions Originating in University Research, 1996–2007*, Final Report. Washington, DC: Biotechnology Industry Organization.

Senoo D., Fukushima M., Yoneyama S. and Watanabe T. (2009) Strategic diversity in Japanese university Technology Licensing Offices. *International Journal of Knowledge Management Studies*, 3, 60–78.

Stercks S. (2011) Patenting and licensing of university research: promoting innovation or undermining academic values? *Science and Engineering Ethics*, 17, 45–64.

Sweeney M. (2012) Correcting Bayh–Dole's inefficiencies for the taxpayer. *Northwestern Journal of Technology and Intellectual Property*, 10, 295–312.

Thursby J., Jensen R. and Thursby M. (2001) Objectives, characteristics and outcomes of university licensing: a survey of major US universities. *Journal of Technology Transfer*, 26, 59–72.

8 | Openness, open innovation à la Chesbrough and intellectual property rights

RENE CARRAZ, ICHIRO NAKAYAMA AND
YUKO HARAYAMA

8.1 Introduction

Intellectual property rights (IPRs) influence innovation and play an important role in diffusing knowledge and creating value (OECD, 2010). By definition, an IPR is an exclusive right designed to encourage private investment in innovation by enabling inventors to recover their investment costs. In the case of a patent, an exclusive right is granted to the inventor to preclude any third party's unauthorized use of her invention and thus personally make exclusive use of it. At the same time, the exclusive right is granted to the inventor as compensation for public disclosure of her invention. In both cases, it is expected that a patent right plays a major part in enabling exclusive use of an invention. These rights assist inventors in getting private returns for their innovation-related investments. In this setting, any free revealing or uncompensated 'spillover' of proprietary knowledge should reduce the innovator's profit from its investments.

Alternatively to this traditional view of the link between innovation and IPRs, new practices for inducing innovation, based on a more collective and open endeavour where exclusivity is not a prerequisite to investment into innovation, are gaining ground. Turning to the innovation strategy of private firms, open innovation has come under the spotlight in recent years. Indeed in a complex technological landscape, it is difficult for a firm to control, investigate and manage all critical constituencies of the innovation process. Thus it is necessary for a firm to find partners to cooperate with, leaving its borders more permeable to knowledge inflows and outflows. As a result, under this open innovation framework, a firm pursuing a profit from innovation may not necessarily stick to exclusive use of its intellectual property

This chapter is adapted from Nakayama and Harayama (2008) and Nakayama (2010).

209

(IP) and, under some circumstances, it may happen that sharing IP without any charge would be a preferred strategy.

On the ground, open innovation is attracting a great deal of interest among US firms, and is spreading to European and Asian firms. Chesbrough (2003, 2006a) cited advanced efforts undertaken by several US corporations (including IBM, Intel, Lucent Technologies, P&G, Air Products, etc.) as examples of open innovation. Murofushi (2008) and Sone (2008) signalled that Japanese firms, albeit on a limited scale, are also seeking to move toward open innovation. In Europe, Rohrbeck *et al.* (2009) have shown how changes in the management process of Deutsch Telecom towards a more 'open' process have enhanced its innovation capacity by opening up its traditional development process and embracing external creativity and knowledge resources. Other case studies and empirical works have been conducted to analyse open innovation practices mainly among European firms (Kirschbaum, 2005; Laursen and Salter, 2006; Van de Meer, 2007; OECD, 2008; EC, 2012). Also, new knowledge exchange mechanisms are emerging, as illustrated by the Eco-Patent Commons, established by World Business Council for Sustainable Development (WVCSD) in 2008, where member firms donate environment-related patents into a pool, making patents available royalty free.[1]

Under open innovation, a high value is no longer attached to the notion of exclusivity that has traditionally been the case, as illustrated by the original motive of patent right. At first glance, one may interpret these phenomena as incompatible with the current IP system, the latter being surpassed by the evolving practices of innovation. The purpose of this chapter is to bring elements of response to this inquiry, with a particular focus on the concept of open innovation developed by Henry Chesbrough, which we call hereafter 'open innovation à la Chesbrough'. To begin with, we attempt to clarify the notion of 'openness', illustrated with the model of 'open science' and 'open source' (Section 2). We then focus on open innovation à la Chesbrough: we analyse the relationship between open innovation and IPRs (Section 3). Finally, we attempt to identify the major impacts of new approach to IPRs on the innovation system (Section 4), and then give our conclusions (Section 5).

---

[1] For Eco-Patent Commons, see the website www.wbcsd.org/web/epc.

Given that the discussions about open innovation are mainly concerned with technological innovation, what follows will be focused on the patent system, which is specifically designed to protect technological ideas, as opposed to other rights under the IPR regime, unless otherwise stated.

8.2 What does 'openness' bring to innovation?

'How does openness influence firms' ability to innovate and profit from their inventions?' is a question that plays an important role in recent research on innovation. Related to this research trend, a central question of this chapter is to understand the rationale of firms' choice to disclose openly some of their knowledge, the phenomenon we have observed more frequently in recent years. This question has already been posed by Nelson (1992, p. 65): 'It appears that, contrary to common beliefs, firms do not keep tight controls on all information about their new technology and in some cases they seem actively to divulge information. How come?'

Indeed openly disclosing knowledge may involve some costs; among others, it can provide useful information to potential rivals, or offer valuable information to other firms without being assured of any direct remuneration. On the other hand, open knowledge disclosure may trigger reciprocity, de facto standards, feedbacks from consumers, network effects or may enhance the reputation of the firm that discloses. Before entering into the heart of this debate, we start with a brief overview of the notion of 'openness'.

8.2.1 *Different types of openness in innovation studies*

The concept of innovation has been well analysed in the literature, given that it is at the centre of many debates. In a Schumpeterian way, innovation refers to the first introduction of a new product, process, method or system into the economy. The notion of openness, despite its recent appeal, is fuzzier. Openness is not a binary classification of open versus closed. It therefore needs to be placed on a continuum, ranging from closed to open, covering various degrees of openness. This discussion is partly related to the evolution from a vertically integrated innovation to a more decentralized approach of innovation, and is thus closely coupled to the debate of the transaction cost economics related to the firm's boundary. In that respect, the question 'how firms make decisions on whether to develop innovation internally or partner with external actors' constitutes a central element in the open innovation literature (Dahlander and Gann, 2010). This creates the need to develop new ways of conceptualizing an innovation process, where the premise of firms vertically integrated is no more the norm, as stated by Langlois (2003): 'large vertical integrated organizations are becoming less significant and are joining a richer mix of organizational forms'.

The notions of openness and unrestricted information sharing may be new in the corporate culture and management literature, but among scientists they have long been identified as institutional norms that are critical to scientific progress (Merton, 1973). More recently, free and open source software initiatives have demonstrated that practising the norms of openness and information sharing in a peer-production setting can result in the creation of complex technological products that are competitive in the market (von Hippel, 2005). In the rest of this section, we will address these two phenomena in turn, 'open science' and 'open source', after offering a clarification of how the word 'open' can be comprehended.

Following Lessig (2001), a resource can be qualified as 'open' if: '(1) one can use it without the permission of anyone else; or (2) the permission one needs is granted neutrally'. This definition leads to two versions of the term, a strong and a weak one. In the strong sense, the owner of the resource has no control on who will have access to it, as no permission will be needed to use it. A typical example of such a process is a publication in a scientific journal. The information embedded in a publication is open in the sense that one cannot restrict its access. With respect to knowledge and technology, a piece of knowledge is open if it is available to all, i.e. all interested parties are given access to it (Pénin, 2007). In a weaker sense, one may have to ask permission, but this permission is not granted at the discretion of the owner of the information, but on a neutral basis. This does not imply that an open resource needs to be automatically free of charge, but that the owner is not able to choose arbitrarily to refuse or grant access to the resource. The famous Cohen–Boyer patent on recombinant DNA provides an example of a situation of a technology being open but not free of charge: under the patent protection, access to the invention was not allowed without permission by the owner; yet it was widely licensed at a reasonable fee without discrimination (Feldman *et al.*, 2007). Anybody, if he/she wanted it, could be granted a licence. This technology was therefore open according to our weak sense, although it was not free of charge.

At this stage it is important to stress the distinction between being free of charge (concept of gratuity) and being open. What we want to understand here is why the latter, according to the Lessig's definition, is critical to fostering innovation and growth. The progress of scientific and technological knowledge is a cumulative process, one that stands on the widespread disclosure of research findings, so they can be tested by the scientific community, to be confirmed or discharged if found to be wrong. Newton's famous remark illustrates this perfectly: 'If I have seen farther it is by standing on the shoulders of giants.' This quotation highlights the cooperative and cumulative characteristics of scientific achievement. Furthermore, it can be argued that the more knowledge you have, the higher the probability that you are going to create new knowledge. This comes from the fact that knowledge can be recomposed in many different ways. In economic terms, this property reveals increasing return in use. Again, this open dimension is somehow different from knowledge being accessible without fee. It is possible that a technology is open, being diffused in a non-discretionary manner, but not free of charge. This open dimension as being constitutive to open innovation is clearly acknowledged by von Hippel and von Krogh (2006) who wrote: 'In our view, free [in the sense of open] revealing of product and process designs is a defining characteristic of "open innovation". Free revealing is the feature that makes it possible to have collaborative design in which all can participate – as is famously the case in open source software projects.'

We will now analyse the characteristics of two institutional arrangements in which openness is the central feature, 'open science' and 'open source software'.

8.2.2 Open science

The issue of openness in fundamental, upstream research is not new. The necessity to preserve the openness of fundamental knowledge, into which innovators can tap in order to generate innovation, has been acknowledged by economists and policy-makers for some time. This idea is, for instance, at the heart of the existence of the open science model, a central dimension of which is the openness of scientific knowledge. 'Open science' is and always has been considered a central element of national systems of innovation, and now is a central theme of research as to how it can adapt itself to increasing interaction with industrial partners, and more proprietary models.

David (1998) refers to 'open science' as 'activities supported by the state funding, the patronage of private foundations and carried on today in universities and public (not-for-profit) institutes'. The notion of open science is an institutional arrangement with an essential collective character, based on an ethos of cooperative inquiry and the free sharing of results. Its comparative efficiency lies in the advantage of open inquiry and complete disclosure of research findings and methods as a basis for the cooperative, cumulative generation of reliable additions to the stock of knowledge. The openness reduces excessive duplication of research efforts, while wide sharing of information places knowledge in the hands of scientists who put them to use. This has the effect of enlarging the opportunities to exploit complementarity within the stock of knowledge, and promoting positive spillovers across research programmes.

Dasgupta and David (1994) argued that the 'open science' arrangement has two fundamental and original economic properties that contribute to its efficiency. First, scientists are those most able to carry out validation and evaluation of their work, in peer-review procedures. They are themselves setting research agendas and evaluating each other's work, hence avoiding principal–agent problems between funding agencies and the research community. Second, since it is the very action of disclosing knowledge that induces the rewards, it creates simultaneous incentives for both knowledge creation and its broad dissemination within the community. On top of that, treating new findings as part of the public domain fully exploits the public goods properties that permit data and information to be concurrently shared in use and reused indefinitely, promoting faster growth of the knowledge stock. Yet the recent trend is for public research organizations increasingly to patent their research results, and this in turn influences their incentive to freely reveal their results. As a result, academic researchers have become more strategic in their choice of what information to disclose in their publications to avoid the possibility of a future patent application being compromised (Webster and Packer, 1997). Concerned by this trend, many voices have advocated the necessity to keep science open, to preserve access to this common good in order to foster the pace of innovation. Fragmentation and appropriation of the scientific common may indeed increase the cost of accessing it and therefore impede the development of follow-on innovations.

8.2.3 Open source

The recent surge of literature about open innovation finds its roots to some extent in the success of free–libre open source software (FLOSS). Indeed, the importance of openness in fostering innovation and the emergence of novelty is well exemplified by this example, which has been extensively analysed in the economic literature (Lerner and Tirole, 2001; Dalle and Jullien, 2003).

The label 'open source' originates from the characteristics of most software. Software is composed of a machine-readable 'object code'; usually software is not written in this format but in a programming language whose 'source code' scripts can be executed by a computer. Most commercial software is distributed only in 'object code' format. The open source movement generally refuses the practice of keeping source code secret, holding instead that source codes should be open and accessible (Boettiger and Burk, 2004). Open source software has two distinct features. First, open source software comes equipped with licences that give existing and future users the right to use, modify and distribute modified and unmodified software to others (Raymond, 1999). Second, it displays a very specific development process. Open source software projects are typically initiated by a 'project leader' who plays a central role in initiating and coordinating the projects while keeping the production of codes by volunteers as decentralized as possible. Depending on their interest in the project, volunteers (or companies) join in and contribute to designing, testing, distributing, and documenting the software. Depending on their knowledge, these voluntary 'project contributors' perform tasks ranging from support, via administration and coordination, to technical development. These contributors, in turn, provide feedback to the open source software developers, share their ideas, report software bugs, indicate new opportunities for using the software, etc. (Raymond, 1999; Lerner and Tirole, 2002). It is a very cooperative and decentralized process, where participants in FLOSS projects intensively interact and exchange information, so that the projects are rapidly designed and debugged.

Open source projects have developed legal instruments to guarantee that the communally produced codes remains freely available and are not captured in proprietary forms. Open source software may be distributed in a variety of licence conditions, but they all have in common the requirement that the recipient of the licensed software is to be provided with the source code. The majority of such licences require licensees who modify or improve the software to make the modifications available on the same terms as the initial software was licenced. Such licences are sometimes termed 'copyleft' licences to indicate what proponents of the open source movement view as a fundamental difference from the system of copyright (Boettiger and Burk, 2004). For instance, in order to ensure that everybody can access the source code and modify and improve software without having to ask for permission from an 'owner', the Free Software Foundation developed an original exploitation licence: the General Public Licence (GPL), the first widely used 'copyleft' licence. The GPL ensures that everybody can use, modify, copy and even distribute software 'protected' by the licence on the sole condition that these changes are kept under the same regime, which means that improvements must remain accessible and free for modifications by everybody. Under this arrangement, everyone has to have free access to the program but it is protected from becoming someone's private IP. The system does not lie outside of the copyright system, it rather uses it in a different and new way to attain its goal of leaving it as open and cooperative as possible to favour cumulative innovation (Lerner and Tirole, 2001).

The success of open source software tends to demonstrate that openness is a sustainable strategy and can foster innovation in some cases. Open innovation frameworks in which competitors share knowledge and information are not specific to software. They have always existed and have often proved to be efficient. This phenomenon is not new. Using historical accounts of the nineteenth-century English blast furnace industry, Allen (1983) stressed that some innovators publicly revealed data on their furnace design and performance in professional society meetings and in published communications. He describes these behaviours as being part of a 'collective innovation' process; this has been illustrated in other historical contexts such as the mining industry during the Industrial Revolution (Nuvolary, 2004). More recently, Henderson and colleagues (Cockburn and Henderson, 1998; Henderson et al., 1999) have highlighted the increasingly frequent publication of research results by some firms. Contemporary examples, such as the

sequencing of the human genome, also support this view of collaborative and open mechanism to support the creation of new scientific and
technological knowledge (McElheny, 2010). Similarly, open source or
free–libre biotechnology is an attempt to transpose the open source
model to biology (Burk, 2002).

8.3 Open innovation à la Chesbrough and intellectual property rights

After the discussion on the concept of 'openness', this section aims to
define the concept of open innovation à la Chesbrough and to analyse
the relationship between open innovation and IPRs.[2] It can be stated,
as we shall soon demonstrate, that Chesbrough's view of openness is
more restrictive in terms of the diffusion of knowledge than the previous constructions we have explained, the open science setting and open
source software. His definition implies that there are actions of
targeted knowledge disclosure, but its access can nevertheless be controlled by each stakeholder depending on the strategic goals of the
firms. Creation of value by setting up a business model and by integrating internal and external knowledge to the firm is central in his
model and there, as noted by Pénin (2008), 'knowledge is not available
to all. It flows only within closed circuit.' The openness here puts
emphasis on the distributive nature of innovation among a wide range
of heterogeneous stakeholders rather than uncontrolled access to it,
which may generate new perception and use of IPRs.

8.3.1 *A changing paradigm*

The concept of 'open innovation' has been introduced and put forward
by Chesbrough (2003, 2006a). In his view, 'open innovation is a
paradigm that assumes that firms can and should use external ideas
as well as internal ideas, and internal and external paths to market, as
firms look to advance their technology' (Chesbrough, 2003, p. xxiv).
This definition is the most commonly used in the management literature. It underscores the fact that firms may take advantage of valuable
ideas and technologies developed outside the firm for their internal

---

[2] From here, when using the term 'open innovation', we will refer, unless otherwise
stated, to the definition introduced by Chesbrough (see above, p. 217).

use, or the other way round: they may make ideas and technologies internally developed available to outsiders.

The open innovation paradigm contrasts with the more traditional practice of innovation, where, from R&D to product sales, all activities are vertically integrated within the firm, a process that Chesbrough described as 'closed innovation'. In the closed innovation context, R&D activities are subject to economies of scale, with the central research laboratory playing an important role. However, the firm's own central research laboratory seems to lose its primacy over time. Rosenbloom and Spencer (1996) have illustrated how leading central research laboratories in the USA were in a dire state, and in the 1980s and 1990s were downsized, redirected and restructured, concluding that this model was 'the end of an era'. In the USA in 1981, 70.1% of all R&D spending was conducted by large companies with 25,000 or more employees. This share decreased to 39.4% in 2001 (Chesbrough, 2006a). Given such a trend, firms seeking innovation are more likely to exploit ideas available outside. Open innovation is an approach that could just meet these needs.

Indeed, in an environment where technologies grow in complexity, demands evolve and diversify rapidly and value chains develop at a global scale, it becomes increasingly difficult to fulfil in-house all competencies needed to ensure all phases of the innovation process. As a result, having access to external knowledge and partners became crucial for firms to maintain their innovation capacity. This is often referred to as 'Joy's law', from Sun Microsystems cofounder Bill Joy, who remarked: 'No matter who you are, most of the smartest people work for someone else' (Lakhani and Panetta, 2007). This new avenue for innovation is illustrated by Chesbrough (2006a, pp. 2–3) as 'there are many ways for ideas to flow into the process, and many ways for it to flow into the market.' The paradigm shift from closed innovation to open innovation leads to greater emphasis on horizontal cooperations, mobility of highly skilled workers, knowledge network and markets, and global value chains, leaving behind the Not Invented Here (NIH) syndrome.

8.3.2 Intellectual property management

The shift to open innovation prompts firms to review their ways of managing IP. Under closed innovation, firms have centred their IP management on pursuing rights to their own research results and using such rights as a means of excluding others or, as a foothold if the right of another party is enforced against them, given that IP created internally will be used to develop their own product. As Chesbrough (2006a, p. xx) puts it plainly, 'if you want something done right, you've got to do it yourself'.

In contrast, open innovation encourages firms to make active use of licensing and transfer of their patents or to reduce information barriers to the public in order to facilitate inflow and outflow of technologies, including external use of unused patents and other IP. Management of IP puts greater emphasis on the marketability of IP rather than its right-to-exclude. Strong IPRs, and patents in particular, are fundamental in ensuring the rise of open innovation. Indeed, firms are more likely to be willing to collaborate and to exchange knowledge if they are protected. Strong IPRs are necessary to secure transactions and exchanges on markets for technologies as they prevent the appearance of free-riding behaviours (Pénin *et al.*, 2011). As a result, open innovation does not require that IP should be opened or released under all circumstances; conversely strong IPR enforcements and the resulting emergence of a market for technology are crucial to the development of the open innovation paradigm.

Chesbrough (2006a) advocates IP management based on the technology life cycle. The latter, inspired by Utterback-Abernathy's product life cycle, is composed of four stages: early stage, growth stage, maturation stage and decline stage. He argued that the way in which IP is managed should be differentiated according to the stage of the product life cycle. For example, in the early stage where neither the marketplace exists nor is a business model in place for an emerging technology, protecting the technology to secure exclusive use has a limited effect on the advent of innovation. Thus more room should be made for sharing information. Exploiting the possibility for potential value creation prevails by getting technological information open to the public. Moving forward in the product life cycle, protection of IP should be strengthened in the course of the technology achieving the dominant market position in terms of design. In the maturation stage, efforts should be made proactively to apply the IP to other industrial fields. And finally, in the decline stage, efforts should be solely directed toward value capture from IP protection.

Chesbrough (2006a) illustrates this argument through the Chinese piracy case of Microsoft Windows. In the Chinese market, Windows and Linux are competing for dominance, and as long as the situation remains unchanged, Microsoft should welcome pirated versions of Windows. According to him, the installation of a pirated version of Windows on a personal computer would preclude installation of Linux and expand market opportunities for complementary products such as applications. If instead great efforts are made to exclude pirated editions, Microsoft may indeed win that battle but at the cost of losing the war for the position of dominance. However, leaving those pirated editions unattended in China is likely to serve as a bad precedent in other regions. Even taking this into consideration, he says, priority should be placed on achieving a dominant position in the Chinese market, until it attains a more mature stage

To make IP management operational under open innovation context, e.g. active use of licensing and transfer of IP rights, the prerequisite is the presence and good functioning of a knowledge market or an intermediate market for innovation wherein IP is traded. Although such a market is in its infancy, the increasing number of transferred patents has demonstrated that the knowledge market is actually growing (Chesbrough, 2006a). On that matter, recent trends show signs of the emergence of a secondary market for IP in the USA and Japan, as well as signs of the development of a small number of intermediary firms specialized in brokering IP resources (Chesbrough, 2006b).

8.3.3 Role of intellectual property rights in open innovation à la Chesbrough

This brief overview of open innovation leads us to recognize that open innovation does not render the IP system obsolete; on the contrary, the former is built upon the latter. With the rise of open innovation, however, IP is expected to fulfil a different role from that which it has traditionally been assigned. Consequently, open innovation requires firms to adapt their IP management to this new avenue.

Behind the debate on IPRs and open innovation, we often observe certain confusion about the interpretation and usage of the term 'open'. For example, the definition of Lessig (2001) ('one can use it without the permission of anyone else') implies unrestricted access to IPR, which goes against the notion of appropriability and restrictive use of IPRs. In the context of open science and open source software, individuals involved act following the norms of unrestricted sharing and releasing of knowledge by putting them in the public domain, with minimal restrictions on those who may be the recipients. The definition of Chesbrough, however, is more restrictive and leaves room for strategic behaviour of knowledge retention, in the form of patents, alliances and joint ventures. In this respect, we can affirm that open innovation à la Chesbrough is not incompatible with the present IP system. In fact, the latter constitutes a basic premise to make open innovation à la Chesbrough viable.

Seen as a business model, open innovation does not preclude the need for an IP strategy. It rather builds on it. Indeed, the open innovation business model pursues a division of innovation labour in an environment in which knowledge is widely dispersed. In that process, it is vital for a firm to integrate internal and external knowledge. This involves two patterns of knowledge flow: an *outside-in* flow that involves the import of useful ideas and technologies that are available outside the company, and an *inside-out* flow that involves the export of unused ideas and technologies to anyone outside the company to make better use of them. As noted by Pénin *et al.* (2011), the outside-in flow is clearly not a new phenomenon; practitioners and researchers have long understood the importance for a firm to monitor and rely on external knowledge, as emphasized by the abundant literature on absorptive capacity (Cohen and Levinthal, 1989, 1990). What is new is the emphasis by the open innovation model of inside-out flows. In contrast to traditional theories, which consider innovation as a core activity that should not be shared or sold, open innovation strongly advocates using external paths to markets, using licences, creating spin-offs and more generally favouring the presence in a market for technology.

It should be also restated here that the term 'open', as used in the context of public opening of IP, does not necessarily mean 'free of charge'. Naturally, in some cases, the term 'open' does mean 'free of charge', as is the case of royalty-free patents release into the public domain. But public opening of patents is a strategy a firm employs for value creation and value capturing, as a part of overall profit maximization. In contrast, as discussed earlier, under open innovation, the active use of licensing and transfer of IP rights is encouraged and the term 'open' here means that a firm, as the owner of the relevant IP, is willing to utilize a third party's technology and let a third party utilize its technology. This departure from the NIH syndrome both in the creation and utilization of technology is premised on a patent right being traded for value.

In this context, an essential prerequisite is that a patent right has been created as a tradeable property right and that a trading rule applicable to the deal has been established under the framework of patent law, which leads us to conclude that open innovation à la Chesbrough cannot properly exist without the IP system and a market for technology.

8.4 Institutional challenges

In the previous section, we argued that open innovation does not render the IP system obsolete. On the contrary, the former is built upon the latter. With the rise of open innovation, however, the role of IPRs is shifting from one traditionally assigned, that is 'the right to exclude any third party', to one 'facilitating knowledge transfer and sharing among stakeholders'. Consequently, open innovation requires firms to adapt their IP management to exploit these new avenues. Does it require a new framework for innovation systems? In other words, do any institutional challenges exist in pursuit of this new practice of IPR management as called for by open innovation? This section formulates some preliminary answers to this question.

8.4.1 Do any institutional challenges exist?

Institutional challenges may come from the fact that open innovation gives more weight to the tradeable property aspect of IP than its characteristic of exclusivity. As mentioned earlier, an essential prerequisite for trading an IPR under an open innovation regime is that an IPR has been created as a tradeable property right, and this under the framework of IP legislation. In other words, the latter shapes the rules of the game affecting transactions of IP. Recognizing that the patent system has the effect of facilitating market transactions by reducing their costs, as shown by the transaction cost theory (Heald, 2005; Merges, 2005), the question arises as to whether or not the present rules and scheme governing transactions of IP are effective in facilitating them.

A related concern is derived from open innovation business models that differ from a more traditional use of IPRs, one based on their exclusive use. The conventional role of a patent is to give a right holder the power to use exclusively its IP. By exercising this power, vertically integrated companies managed to control the market of their patented products in a close innovation model. Legally, remedies for infringements lay the foundation of the exclusivity of IPR. In particular, injunctions enable patentees to effectively exclude the competing infringers from the market, by directly stopping their ongoing acts of infringements (e.g. by stopping the production of incriminated product) and preventing future infringements. In addition, patentees may demand damage reparations. The main purpose of damage compensation is, however, a monetary compensation for past infringements. Even though it may deter, to some extent, infringements and contribute indirectly to the exclusivity of a patented invention, it is only an indirect process as compared to injunctions which play a more critical role in protecting exclusivity.

Open innovation encourages right holders, through licensing or other IP transactions, to share technologies with external partners in a dynamic process of collaboration. In that setting, the IPR system functions as a vehicle for technology transfer, by creating a business model in which a right holder does not necessarily exclusively use their invention. This usage of the IPR system may raise the question about the necessity of granting an exclusive right to intellectual creations in the first place. Indeed, if right holders of patented inventions no longer persist in exclusive use of their IPR, is there any reason to grant an exclusive usage to intellectual creations which, by nature, are available for parallel use by any third party? In addition, emerging IP markets under open innovation may give rise to the concern that a patent may come into the hands of those who abusively exercise the exclusive power, as discussed later in this section in the case of 'patent trolls'. Given that injunctive relief is at the heart of the exclusive right of IP, these changes might make it necessary to reconsider the modality of injunctive relief.

It should be further noted that open innovation would have a wider variety of players involved, compared to the traditional closed innovation model founded on vertical integration of the research process. Among them, universities are becoming privileged partners as providers of new ideas and technologies. By definition, universities are repositories and drivers of science, and have the status of non-profit organizations. Entering into research collaboration and transactions with private firms, in particular while being in a position to negotiate with them over IP issues, implies that universities need a clear rule defining the ownership of IPs created by their constituents, including students, and rules for sharing the cost and distributing the revenues generated by IP transactions. Also, this partnership may lead universities to depart from the principle of open science and one of its core elements, the unrestricted diffusion of research results.

In sum, we recognize that open innovation may raise some institutional challenges for existing innovation system. In what follows, we will focus on these three points: the development of the IP market, modality of injunctive relief, and universities as partners.

8.4.2 Development of the intellectual property market

Under open innovation, the volume of IP transactions is expected to increase. Therefore the presence of a well-functioning IP market becomes a critical underlying condition. Indeed, it can be argued that no division of innovative labour is possible without markets for technology (Arora *et al.*, 2001). In that line of thought, recent studies from the OECD have stressed the importance of knowledge networks and markets as an enabler of new business models that take advantage of the greater reliance on outsourcing and insourcing of R&D (OECD, 2012). But given the novelty of such markets, there is a need for further research on their structure, dynamic and efficiency. For instance in the case of an IP transaction, it often occurs as a result of negotiation between two parties, through which they decide the terms of transactions, respecting the rules of the game fixed by the patent system; however, it does not necessarily mean that such transactions take place widely on the IP market. In fact, bilateral negotiation or intra-firm trade remain privileged channels for IP transactions. For a market that brings underutilized invention to potential buyers to emerge, tools to evaluate, report and trade IP need to be developed.

Additionally, the present situation of an undermobilized patent market is partly attributable to a lack of substantial need for right holders to trade their IP, reflecting the traditional dominance of players operating within a vertically integrated organization. However, the spread of open innovation will naturally increase the number of transactions in the IP market, and no further consideration would be required with respect to institutional adjustment, if there is no major market failure.

However, a fundamental problem arises when assessing the price and negotiating the deal. Indeed, firms often encounter difficulties in evaluating the value of invention, and, practically, the terms and conditions of the transfer contract or licence agreement are negotiated on a case-by-case basis in most cases. It increases transaction costs, which constitutes a barrier for the development of an IP market.

Nevertheless, a recent report by the World Intellectual Property Office (WIPO) points out the rise of an IP-based knowledge market (WIPO, 2011). It shows that international royalty and licensing payments and receipts are growing, and that their growth outpaces the growth in global GDP. However, it should be noted that the data include intra-firm payments. In addition, fewer data are available on domestic IP transactions. Recognizing data limitation in measuring the phenomenon, WIPO (2011) still suggests that the IP market is on the rise, although it starts from low initial levels and therefore is still relatively small.

Chesbrough is optimistic about the future growth of the IP market, noting the presence of agents acting as intermediaries to operate this market. Through their practices, they may gain professional expertise in valuation of inventions. Given the high frequency of unused patents, an increase in the number of these intermediate agents could be considered as a remedy to this problem of IP market. At the same time, however, the increasing importance of intermediate agents may have a bearing on eventual patent trolls, given that an intermediate agent does not utilize a patented invention himself, and consequently his interest resides in the transaction itself and not in the use of invention. There is concern that the increasing number of intermediate agents facilitates the existence of patent trolls.

8.4.3 Patent trolls: modality of injunctive relief

Although no clear definition exists, the term 'patent troll' is generally used to mean a party who, with no intent of utilizing patented inventions, purchases patents from firms, often in a state of bankruptcy, and imposes outrageous licensing fees on firms who are utilizing the patented invention(s) under the threat of injunction. Thus, the term patent troll refers to patent holders who try to provoke hold-up situations, i.e. who use the threat of an injunction to extract excessive value far beyond the true economic value of the invention, taking advantage of the sunk cost the infringers have to bear. They should be differentiated from non-practising entities (NPE) such as universities or public research institutions that are actively engaged in technology transfer. They develop technology by themselves or play a role between technological firms and manufacturing companies on markets for technology. Their aim is to transfer technologies through the license of the IPRs they manage. On the other hand, patent trolls keep their patent portfolio hidden and seek to be infringed (Reitzig *et al.*, 2010). Trolls are not engaged in licensing activities for technology transfer, but instead they speculate on patent litigation. Normally, in cases where a patent dispute arises between parties utilizing a patented invention, they settle by entering into a cross-licence contract to prevent either party's business from being suspended by virtue of the other party's patent right. If the right holder is a patent troll who does not actually utilize the invention, however, it becomes impossible to enter a cross-licence contract between the parties concerned.

In the context of open innovation, Chesbrough (2006a) recognizes the potential risk of patent trolls as a side-effect of the emergence of an active IP market. As a defensive measure for forestalling patent trolls, he suggests pre-emptive buying of patents. In fact, according to news reports, large US firms have established an organization dedicated to buying patents as a form of self-defence.[3] In the USA, a decision of the Supreme Court in 2006 held that, even if the infringement of a patent right is found by the court, it should not automatically grant an injunction but rather determine whether to grant or deny injunctive relief, within its discretion in accordance with equitability, thereby leaving the door open to impose restrictions on the granting of injunctions.[4] This is one example of how, in the USA, where the problem of patent trolls has come under close scrutiny, people appear to have started taking measures against it.

Fundamentally, the current patent law does not explicitly require a patentee to utilize the patented invention. Given this, it ought to have

---

[3] According to the *Wall Street Journal*, 30 June 2008, 'Tech giants join together to head off patent suits', the newly established organization is named Allied Securities Trust. Currently it has 25 members from Europe, North America and Asia, mainly in IT sectors (www.alliedsecuritytrust.com/ASTMembers.aspx). Another example of a defensive patents aggregator is RPX which has 50 members (www.rpxcorp.com/index.cfm?pageid=11).

[4] *eBay Inc.* v. *MercExhange*, 547 US 388 (2006).

been anticipated that a patentee failing to make use of the patented invention can still have his patent right enforced. Under the regime of closed innovation with a predominance of vertically integrated organizations, this issue did not have any reason to be addressed. Simply, the phenomenon of patent trolls is becoming apparent, as the number of players non-vertically integrated has increased with the advent of open innovation.

As we stated earlier, if a business model in which a right holder does not use his own invention exclusively prevails, then the necessity or validity of granting an exclusive right for intellectual creations may be challenged; that is not the case for the time being. Also, the transaction cost theory suggests that the creation of an exclusive property right would also have a promotional effect on the trading of inventions under open innovation. Consequently, it would not be necessary, for the time being, to walk away from the principles underlying the patent protection framework based on the idea of exclusive rights or injunctive relief. However, it is not to say that the exclusive-right or injunctive-relief framework of IP protection should be regarded as an absolute essential (Nakayama, 1996). On the contrary, it would suggest that the focus of discussion should be shifted from whether to limit injunctive relief to in what circumstances such right should be limited, to whether such question should be clarified beforehand in legislation or be decided upon by the courts on a case-by-case basis,[5] and what would be the appropriate amount of compensation for damages in cases where injunctive relief is limited.

8.4.4 Universities as partners

It is increasingly stressed that universities have to contribute actively to their respective national innovation system. They should not only create new scientific and technological knowledge, but they have also to be actively involved in transferring their research results into

---

[5] Or falling somewhere in between, a framework similar to government use (28 U.S.C. §1498) in the USA, for example, is possible in which patentees shall not seek injunction against not only the government that infringed the patent, but also those who infringed it for the government and with its 'authorization or approval' while the court decides on compensation for damages (Nakayama, 2008).

commercial success. The notion of 'entrepreneurial university' or 'academic capitalism' refers to universities being involved in the transfer of their research results through patenting, licensing, collaborative work with industry and more generally a greater involvement in economic and social challenges (Harayama and Carraz, 2012). Following this trend, in line with the rise of the open innovation paradigm, universities have emerged as potential partners vis-à-vis private firms for sourcing knowledge and technologies, accompanied by the recent trend of universities pursuing patent registration and licensing with respect to inventions created through joint efforts by the government and others. This trend seems to be consistent and coherent with the concept of open innovation; however, a careful analysis indicates that it may be a new source of contingency.

By the fact that the purpose of open innovation is to utilize external knowledge, Chesbrough takes a positive view of the promotion of industry–university cooperation, but he remains cautious about the pursuit by universities under the US Bayh–Dole Act of patenting activities with respect to inventions resulting from such cooperation. Chesbrough (2003) argues that kernels for next-generation technologies should be disseminated widely and rapidly, noting that the priority given by universities for patenting and licensing activities, may hinder the dissemination of useful knowledge, in particular in the field of basic research. The same concern is formulated by David (2003) who analyses at length the potential pitfalls of such trends on the open science mechanisms and incentive structure.

The fact is that, in the last 30 years, there has been a radical increase in the number and share of academic patents has been noticed first in the USA, then in Europe and more recently in Japan (Mowery *et al.*, 2001; Geuna and Nesta, 2006; Takahashi and Carraz, 2011). Between 1969 and 1986, universities owned 1.1% of US patents issued, and by 1999 that number had risen to 4.8% (Eisenberg, 2003). In Europe, the share of public research institutions' filings (including universities) in total patent applications at the European Patent Office has jumped from about 0.5% in 1981 up to nearly 4% in the early 2000s (Zeebroeck *et al.*, 2008). It must be kept in mind that these figures are lower-bound measures, as many university-invented patents are assigned to non-academic institutions.

This increase has been induced by an active institutional level engagement of universities in the creation and management of IPRs.

A major symbol of this trend is the legislative frenzy that started with the US Bayh–Dole Act in 1980, and went through subsequent similar provisions in Europe and Japan (Verspagen, 2006; Takahashi and Carraz, 2011). The Act allowed universities to claim ownership of inventions made as a result of federally funded research. The motivation of Congress in passing the Act originated from the proposition that patents resulting from federally funded research were unexploited due to lack of clear rules on their ownership. It assumed that university ownership of faculty inventions facilitates their commercialization, thus enhancing economic efficiency. This institutional change induced an increasing pressure to translate the results of their work into privately appropriable knowledge. In that sense it collides with the free diffusion of research results, a crucial element of the ethos of open-academic science (as we have seen in Section 2). The resulting question is how this limitation of the free flow of academic knowledge, departing for the open science norms, can be offset by the economic gains of increasing pool IPRs to be traded in the knowledge market. The research exemptions could be a response to this problem, but the debate on how to translate them in terms of operational rules, ensuring a right balance between 'research use and patent holder's rights' is far from closed (OECD, 2006).

8.5 Conclusion

In this chapter, after a brief review of the concept of 'openness' and the discussion around three different perspectives of openness, e.g. open science, open sources and open innovation à la Chesbrough, we examined the relationship between open innovation and the IP system, with a particular focus on patent. Then we identified and analysed some major institutional challenges induced by this changing practice of IPR management, namely the development of an IP market, modality of injunctive relief, and universities as partners in the IP market.

As observers have occasionally suggested, the vertically integrated closed innovation model has exposed its own limitations. The significance of open innovation lies in setting out a new model in which, based on the premise of a non-vertically integrated model, equal importance is attached to external knowledge and internal knowledge and efforts are made to create and capture value by combined use of the two, together with other resources. Open innovation is also characterized by its emphasis on the need for firms to have strategic IP management.

By capturing open innovation in their development strategy, firms opt for using the patent system as a means of pursuing profit. In the first place, the term 'public opening' has a strategic and equivocal meaning, which implies that open innovation prompts users of the patent system to further improve their patent management. So, basically, open innovation sends a message specifically to users of the patent system. While conventional wisdom puts a focus on exclusivity of a patent right, open innovation à la Chesbrough urges company managers to reconsider the role of patents and use them as vehicles for technology transfer in IP markets.

Policy-makers should also pay attention to what these changes bring about. The patent system, with the concept of exclusivity at its heart, has been considered consistent with closed innovation for some time. However, open innovation à la Chesbrough makes it clear that patents are tradeable property rights. Keeping this in mind, policy-makers should carefully revisit institutional design to make sure that technology transfer through IP markets contributes effectively to accelerating innovation. More generally open innovation practices entail many opportunities for firms, as well as some threats and costs, leading to new challenges for managers and policy-makers. Firms need to develop practices to deal with external knowledge flows and to build strategies of knowledge integration tailored to different partners and level of openness. Depending on the circumstances and partners, firms should diffuse their knowledge on an unrestricted basis, build long-term cooperation with different actors such as universities, or monetize their inventions on IP markets and networks. Further studies on these topics should investigate how firms should organize their organizational boundaries to fit their innovative strategies and produce guidelines from managers to manipulate different options brought about by a more open and decentralized model of innovation.

References

Allen, R. (1983). Collective inventions. *Journal of Economic Behavior and Organization*, 4, 1–24.

Arora, A., Fosfuri, A. and Gambardella, A. (2001). *Markets for Technology*. Cambridge, MA: MIT Press.

Boettiger, S. and Burk, D. (2004). Open source patenting. *Journal of International Biotechnology Law*, 1, 221–231.

Burk, D. L. (2002). Open source genomics. *Boston University Journal of Science and Technology Law*, 8, 254–260.

Chesbrough, H. (2003). *Open Innovation: The New Imperative for Creating and Profiting from Technology*. Boston, MA: Harvard Business School Press.

(2006a). *Open Business Models: How to Thrive in the New Innovation Landscape*. Boston, MA: Harvard Business School Press.

(2006b). *Emerging Secondary Markets for Intellectual Property: US and Japan Comparisons*. Tokyo, Japan: National Center for Industrial Property Information and Training.

Chesbrough, H., Vanhaverbeke, W. and West, J. (2006). *Open Innovation: Researching a New Paradigm*. Oxford: Oxford University Press.

Cockburn, I. and Henderson, R. (1998). Absorptive capacity, coauthoring behavior, and the organization of research in drug discovery. *Journal of Industrial Economics*, 46, 157–182.

Cohen, W. M. and Levinthal, D. A. (1989). Innovation and learning: the two faces of R&D. *Economic Journal*, 99, 569–596.

(1990). Absorptive capacity: a new perspective on learning and innovation. *Administrative Science Quarterly*, 35, 128–152.

Dahlander, L. and Gann, D. M. (2010). How open is innovation? *Research Policy*, 39, 699–709.

Dalle, J.-M. and Jullien, N. (2003). 'Libre' software: turning fails into institutions? *Research Policy*, 32, 1–11.

Dasgupta, P. and David, P. (1994). Toward a new economics of science. *Research Policy*, 23, 487–521.

David, P. (1998). Common agency contracting and the emergence of 'open science' institutions. *American Economic Review*, 88, 15–21.

(2003). *Can 'Open Science' be Protected from the Evolving Regime of IPR Protections?*, Working Paper 03–011. Stanford, CA: Department of Economics, Stanford University.

Eisenberg, R. (2003). Science and the law: patent swords and shields. *Science*, 299, 1018–1019.

European Commission (EC) (2012) *Open Innovation*. Brussels, Belgium: Directorate-General for the Information Society and Media.

Feldman, M., Colaianni, A. and Liu, C. (2007). Lessons from the commercialization of the Cohen–Boyer patents: the Stanford University Licensing Program. In *Intellectual Property Management in Health and Agricultural Innovation: A Handbook of Best Practices*, A. Krattiger *et al.* (eds.). Oxford: MIHR. www.ipHandbook.org.

Geuna, A. and Nesta, L. (2006). University patenting and its effects on academic research: the emerging European evidence. *Research Policy*, 35, 790–807.

Harayama, Y. and Carraz, R. (2012). Addressing global and social challenges and the role of university. In *Global Sustainability and the
Responsibilities of Universities*, L. E. Weber and J. J. Duderstadt (eds.),
pp. 119–130. London: Economica.

Heald, P. (2005). A transaction cost theory of patent law. *Ohio State Law
Journal*, 66, 473–509.

Henderson, R., Orsenigo, L. and Pisano, G. (1999). The pharmaceutical
industry and the revolution in molecular biology: interactions among
scientific, institutional and organizational change. In *Sources of Industrial Leadership*, D. C. Mowery and R. R. Nelson (eds.), pp. 267–311.
New York: Cambridge University Press.

Hippel, E. von (2005). *Democratizing Innovation*. Cambridge, MA: MIT
Press.

Hippel, E. von and G. von Krogh (2006). Free revealing and the private-
collective model for innovation incentives. *R&D Management*, 36,
291–302.

Kinukawa, M. (2008). *Open Innovation and Voluntary Public Opening of
Research Products*, Research Report 312. Tokyo, Japan: Fujitsu
Research Institute.

Kirschbaum, K. (2005). Open innovation in practice. *Research Technology
Management*, 48, 24–28.

Lakhani, K. and Panetta, J. A. (2007). The principles of distributed innovation. *Innovations*, 2, 97–112.

Langlois, R. (2003). Transaction cost economics in real time. *Industrial and
Corporate Change*, 12, 351–385.

Laursen, K. and Salter, A. (2006). Open for innovation: the role of openness
in explaining innovation performance among UK manufacturing firms.
Strategic Management Journal, 27, 131–150.

Lerner, J. and Tirole, J. (2001). The open source movement: key research
questions. *European Economic Review*, 45, 819–826.

Lessig, L. (2001). *The Future of Ideas*. New York: Vintage Books.

McElheny, V. (2010). *Drawing the Map of Life: Inside the Human Genome
Project*. New York: Basic Books.

Merges, R. (2005). A transactional view of property rights. *Berkeley Technology Law Journal*, 20, 1477–1520.

Merton, R. (1973). *The Sociology of Science: Theoretical and Empirical
Investigations*. Chicago, IL: University of Chicago Press.

Mowery, D., Nelson, R., Sampat, B. and Ziedonis, A. (2001). The growth of
patenting and licensing by US universities: an assessment of the effects of
the Bayh–Dole Act of 1980. *Research Policy*, 30, 99–119.

Murofushi, Y. (2008). Open innovation in pharmaceutical industries. *Patent
Research*, 46, 19–26.

Nakayama, I. (2008). Some observations about the balance between protection and use of and exclusivity of a patent right. In *Intellectual Property Policy and Management*, K. Sumikura (ed.), pp. 211–232. Tokyo, Japan: Hakuto-Shobo Publishing.

(2010). Open innovation and the patent system. *Annual Report of Japan Association of Industrial Property Law*, 33, 135–160.

Nakayama, I. and Harayama, Y. (2008). Open innovation and intellectual property. *Patent Research*, 46, 6–18.

Nakayama, N. (1996). *Multimedia and Copyright*. Tokyo, Japan: Iwanami-Shoten.

Nelson, R. (1992). What is commercial and what is public about technology and what should be? In *Technology and the Wealth of Nations*, N. Rosenberg, R. Landau and D. Mowery (eds.), pp. 57–71. Stanford, CA: Stanford University Press.

(2005). Linkages between the market economy and the scientific commons. In *International Public Goods and Transfer of Technology*, K. E. Maskus and J. H. Reichman (eds.), pp. 121–138. Cambridge: Cambridge University Press.

Nuvolari, A. (2004). Collective invention during the British Industrial Revolution: the case of the Cornish pumping engine. *Cambridge Journal of Economics*, 28, 347–363.

OECD (2006). *Research Uses of Patented Knowledge: A Review*, STI Working Paper 2006/2. Paris: OECD.

(2008). *Open Innovation in Global Networks*. Paris: OECD.

(2010). *The OECD Innovation Strategy: Getting a Head Start on Tomorrow*. Paris: OECD.

(2012). *Knowledge Networks and Markets in the Life Sciences*. Paris: OECD.

Pénin, J. (2007). Open knowledge disclosure: an overview of the empirical evidences and the economic motivations. *Journal of Economic Surveys*, 21, 326–348.

(2008). *More Open than Open Innovation? Rethinking the Concept of Openness in Innovation Studies*, Working Paper 2008–18. Strasbourg, France: BETA

Pénin, P., Hussler, C. and Burger-Helmchen, T. (2011). New shapes and new stakes: a portrait of open innovation as a promising phenomenon. *Journal of Innovation Economics*, 1, 11–29.

Raymond, E. (1999). *The Cathedral and the Bazaar*. www.tuxedo.org/-esr/writings/cathedral-bazaar.

Reitzig, M., Henkel, J. and Schneider, F. (2010). Collateral damage for R&D manufacturers: how patent sharks operate in markets for technology. *Industrial and Corporate Change*, 19, 947–967.

Rohrbeck, R., Hölzle, K. and Gemünden, H.G. (2009). Opening up for competitive advantage: how Deutsche Telekom creates an open innovation ecosystem. *R&D Management*, **39**, 420–430.

Rosenbloom, R. and Spencer, W.J. (1996). *Engines of Innovation: US Industrial Research at the End of an Era*. Boston, MA: Harvard Business School Press.

Sone, K. (2008). Open innovation in the automobile industry. *Patent Research*, **46**, 27.

Takahashi, M. and Carraz, R. (2011). Academic patenting in Japan: illustration from a leading Japanese university. In *Academic Entrepreneurship in Asia: The Role and Impact of Universities in National Innovation Systems*, P.K. Wong (eds.), pp. 86–103. Cheltenham, UK: Edward Elgar.

Van de Meer, H. (2007). Open innovation – the Dutch treat: challenges in thinking in business models. *Creativity and Innovation Management*, **6**, 192–202.

Verspagen, B. (2006). University research, intellectual property rights and European innovation systems. *Journal of Economic Surveys*, **20**, 607–632.

Webster, A. and Packer, K. (1997). Patents in public sector research: when worlds collide. In *Universities and the Global Knowledge Economy: A Triple Helix of University–Industry–Government Relations*, H. Etzkowitz and L. Leydesdorff (eds.), pp. 47–59. London: Leicester University Press.

West, J. (2006). Does appropriability enable or retard open innovation? In *Open Innovation: Researching a New Paradigm*, H. Chesbrough, W. Vanhaverbeke and J. West (eds.), pp. 109–133. Oxford: Oxford University Press.

WIPO (2011). *The World Intellectual Property Report 2011: The Changing Face of Innovation*. Geneva, Switzerland: WIPO.

Zeebroeck, N., Van Pottelsberghe, B. and Guellec, D. (2008). Patents and academic research: a state of the art. *Journal of Intellectual Capital*, **9**, 246–263.

Conclusion

JEAN-CLAUDE PRAGER

The global economy is currently experiencing changes that are significant for the future of patents as incentives for innovation. We may focus on four central questions.

Question 1

The first question is about the possible evolution patterns of patent markets. For a better understanding, we should return to the basic forms and functions of markets. A market is an economic subsystem comprising a collection of agents with diverse motivations, but pursuing mainly utilitarian or lucrative goals. These agents are in relation to one another to buy, sell or exchange rights concerning goods and services that have common characteristics. These transactions take place according to processes defined by formal or informal rules, possibly under the authority of an (or many) operator(s) ensuring these rules are respected. The collection of these transactions on the market for a good or service is usually composed of two segments:

- A segment of interpersonal or relational transactions, between identified interlocutors who negotiate directly between them the conditions of the transaction. These transactions happen under various forms, most frequently face-to-face with the assistance of intermediaries. Braudel (1985) refers to these transactions as 'the inferior register, the markets, the shops, the peddlers'.
- A segment of impersonal transactions that involves standardized goods and exchanges taking place in central marketplaces, benefiting from price determination by collective supply-and-demand adjustment mechanisms controlled in an institutional framework. For example, these mechanisms can be continuous or discontinuous

The author would like to thank Simon Lapointe for helpful comments.

auction systems on individual products or lots. This segment corresponds to the superior register of Braudel (1985), the one of fairs and stock exchanges. It also corresponds to the Walrasian model of shared information, and to the modern market economy.

The degree of specificity of exchanged goods and services can be variable; in many cases, the definition of the object of exchange is relatively indeterminate (as in incomplete contracts). The reputation of the seller is one of the most important means to reduce the uncertainty of the quality of the purchased product. For a given category of goods and services, the two principal forms of exchange can coexist, depending on the technical nature of the individual products exchanged and on the needs of operators. Those two forms of exchange can exist for the same categories of goods, such as wines (one can obtain the same bottles at retailers, producers, through negotiation between amateurs, or at public auctions), or such as stocks from publicly traded enterprises, for which transaction blocks can go through negotiations over the counter, under conditions generally different from those determined on the marketplace (mergers and acquisitions).

What we call the development of the market for a category of goods or services thus involves three evolutions different in logic, but complementary:

- An increase in the proportion of market transactions in the set of exchanges on a given category of goods or services (a historic example is the evolution from subsistence agriculture to national and international markets).
- Improvements in the operation of the market through a reduction in its imperfections as defined in classical microeconomics (uncertainties related notably to information asymmetry, externalities, etc.), thus leading to increased collective welfare. In this regard, a reinforcement of the quality of relational contracts and of the common knowledge reduces information asymmetries and possibilities for opportunistic behaviour, in such a way that can be cumulative.
- An enlargement of the market through the growth of global consumption and production of the good. This enlargement is usually historically associated with the division of labour, increased standardization of the good, and economies of scale favouring greater collective efficiency.

Today, knowledge, technologies and patents are strongly related to each of these three dimensions, to an extent that varies with the sector or technology. We can classify technologies according to many criteria: the degree of complexity measured as the number of elementary inventions present on average in each product, the degree of knowledge or skills included, the size of the market for the technology, the possibility of division of labour, the degree of maturity of products that are principally concerned by the technology or can be inversely related to the distance to market of products derived from the technology.

Simple technologies that have a short distance to market and a low level of included knowledge and skills, and that have simpler and mainstream-derived products, are more prone to the break-up of the knowledge value chain and to serve as support for impersonal market transactions. This case is particularly relevant to the sectors of fine chemicals, cosmetics or pharmacy, which will probably experience in the upcoming years growth in intellectual property markets, with exclusion and licensing strategies increasingly prominent. In other sectors, according to the complexity of the chain of value of knowledge, strategies involving the purchase of patents and other intellectual property assets can be varied. In the information and communication technologies (ICT) industries, in which numerous patents cover every final product (many thousands for a mobile phone), the management of patents (including transactions) is done directly on entire portfolios, in bulk, aiming for the reinforcement of the competitive position of the parties involved (Harhoff *et al.*, 2008). In contrast, technologies that are closely related to specific industrial systems, or strongly integrated, are less prone to the impersonal circulation of rights that are associated with them. The evolution of patent markets will thus probably take very diverse paths according to the related technologies. Indeed, some technologies invite a better definition of intellectual property rights, while others instead favour a reduction of those rights, if not an outright elimination.

Question 2

The second question is about intellectual property rights at an international level, especially those concerning global public goods such as health and the environment. This subject has been of importance for the past ten years or so. How can we reconcile the pursuit of innovation in developed countries and the needs of emerging economies? This question is particularly relevant for pharmaceutical innovations or green technologies. In the case of pharmaceutical patents, a solution was found in international trade agreements through the introduction of a mechanism of flexibility allowing Member States to deliver compulsory licences in national emergency situations, in cases of public health crises, such as those related to AIDS or other epidemics. The compulsory licence is a pressure tactic convincing patent holders to decrease the price of their products or the price of a licence, with low royalty rates. The situation has improved considerably since the Doha Agreements of 2001, which authorized countries that do not produce pharmaceuticals to purchase them from other developing countries that do, under compulsory licences. However, patents do not provide sufficient incentives to engage in further R&D on drugs for diseases that do not constitute an important market in the developed world. If individuals affected by these diseases or their governments do not have the resources to pay prices sufficiently higher than the marginal cost of production of the treatment, firms will not be able to amortize the fixed costs of research, independently of the level of protection granted by patents (Kyle and McGahan, 2009).

In the case of green technologies, the goal of the reduction in greenhouse gas emissions constitutes a global public good: the fight against global warming. Developing countries wish to obtain exemptions from intellectual property rights. Ideas proposed include the suspension of patents in favour of less advanced countries, a reduction in the length of protection, the creation of compulsory licences similar to those in the pharmaceutical industry, and the creation of a multilateral fund for climate technologies that would buy the most sensitive technologies to put them at the disposal of developing countries. The Copenhagen Agreement of 2009 included provisions for the creation of a technological mechanism 'responsible for speeding up the development and the transfer of technologies supporting adaption and attenuation measures' to allow the purchase of certain patents and grant free licences to developing countries.

Question 3

More generally, we can ask the question whether patents are actually useful for technical progress and growth.

That question is not recent and was formally posed by E. Penrose in 1951, then by F. Machlup in 1958 (Penrose, 1951; Machlup, 1958). According to these two authors, there would be just as little justification for creating a system of intellectual protection today if it did not exist, as there is to eliminate the one that currently exists. Since then, the debate has evolved, notably in light of the explosion of patents since the 1950s and of numerous empirical studies. Periods of accelerated innovation are generally associated with increased patent activity (for example, electricity in the 1880s and ICT in the 1990s). However, the causality is not only from innovation to patents, because the latter can also become the source of rents, thus fuelling rent-seeking behaviour (Bessen and Meurer, 2008). The base case of the utility of a patent for innovation is the one in which the patent corresponds to a new well-identified product, for which knowledge is isolated (in other words, that is not part of a chain). This form of invention is more frequent in some sectors, such as pharmacy. In other situations, the effects of patents are more complex (Scotchmer, 2004; Hall, 2007; Bessen and Maskin, 2009). Indeed, economic studies show that a reinforcement of intellectual property is not systematically accompanied by an increase in the number of patents induced (Hunt, 2006; Allred and Park, 2007). Overall, four major conclusions emerge (Hall, 2007):

- First, introducing or strengthening a patent system (lengthening the patent term, broadening subject matter coverage or available scope, improving enforcement) unambiguously results in an increase in patenting and also in the use of patents as a tool of firm strategy...
- It is much less clear that these changes result in an increase in innovative activity...
- If there is an increase in innovation due to patents, it is likely to be centred in the pharmaceutical, biotechnology, and medical instrument areas, and possibly specialty chemicals...
- Finally, the existence and strength of the patent system affects the organization of industry, by allowing trade in knowledge, which facilitates the vertical disintegration of knowledge-based activity.

Boldrin and Levine (2008) make an even stronger case, challenging the role of patents. The authors remind us that all great discoveries were made by scientists not concerned by material rents from innovation, and that Watt's quest for an intellectual monopoly rent probably harmed the early development of machinery and of scientific and technological progress more generally. Furthermore, litigation costs have greatly increased in the past two decades, such that the patent system may especially be of benefit to lawyers. Indeed, in the United States, litigation costs are greater than revenues from intellectual property, except for the chemical and pharmaceutical sectors (Bessen and Meurer, 2008). In theory, Boldrin and Levine (2008) argue that the conventional layout of creative activities with high fixed costs and zero marginal costs is debatable and that the quasi-rents already present in a perfectly competitive economy are sufficient to motivate innovation. They also remark that patents only serve to protect yesterday's innovators from new innovations, and that the patent system creates an unnecessary monopolistic distortion. The authors then conclude that patent law should be progressively dismantled and replaced by subsidies for R&D, or by prizes awarded by state authorities to inventors. The argument has been strongly opposed, but it forces the reassessment of the positive effect of patents on innovation and the evolution of a judicial system that presents many inconveniences. To what extent is the patent a sub-product as well as a factor of innovation (Aghion, 2010)?

Question 4

The final question is normative, and is related to the historical tendency of the development of market economies. The perspective of an ideal knowledge economy in which the access to and usage of knowledge are secured by transparent and equitable market transactions, or a democracy of knowledge, is misguided in light of the historical realities of polarization, inequalities and instabilities that characterize immaterial goods at least as much as material ones. This perspective also disregards the collective will to accept the commoditization of knowledge in general, because the latter is closely related to human nature and cannot be detached from it (see, for example, Arrow, 1997). In this matter, we have to ask about the equilibrium between the three possible economic regulation mechanisms: public management, benevolence (altruism in individual utility functions) and economic incentives.

Extremes are equally problematic, even if the debate is limited to considerations of economic efficiency. Regulation through pure benevolence remains an unachievable ideal and assumes an international consensus that is impossible given the variety of interests represented. One need only note that consensus is already impossible even for obvious global public goods. This type of regulation also supposes a partitioning harmful to diffusion of innovations. Regulation with a bureaucratic and non-competitive system of research and valorization has reached its limits, given the entropic reality of modern innovation. The trend of markets, in which the portion of benevolence and of relations decreases progressively, is also undesirable given that a strong relational and cooperation component is necessary for the smooth operation of knowledge networks. And economic incentives do not take into account the 'intrinsic motivation of researchers', and 'explicit incentive schemes may sometimes backfire, especially in the long run, by undermining agents' confidence in their own abilities or in the value of the rewarded task' (Benabou and Tirole, 2003). Challenges are numerous: how can modern ethical requirements be integrated to the knowledge economy, while still optimizing incentive regimes? How can we ensure that markets develop in a competitive way, to stop the emergence of global knowledge monopolies? How can we make sure that the development of market mechanisms does not happen in a way that is detrimental to the development of non-market mechanisms that are sometimes more efficient in the transfer of knowledge, and in a way that does not impede the global effort in fundamental research in a period of rarity of public resources? How can patent markets be designed in a way that improves the insertion of small and medium enterprises in the knowledge circuits instead of subordinating them? How can these markets be designed so they do not serve as a lever in the extortion of rents with the help of dubious patents? These challenges are often made more complex by their global scope and the variety of national cultures.

References

Aghion, P. (2010) *Commentaire: Les marchés de brevets dans l'économie de la connaissance*. Paris: Conseil d'analyse économique, La Documentation Française.

Allred, B. B. and W. G. Park (2007) The influence of patent protection on firm innovation investment in manufacturing industries. *Journal of International Management*, 13, 91–109.

Arrow, K. (1997) Invaluable goods. *Journal of Economic Literature*, 35, 757–765.

Benabou, R. and J. Tirole (2003) Intrinsic and extrinsic motivation. *Review of Economic Studies*, **70**, 489–520.

Bessen, J. and E. Maskin (2009) Sequential innovation, patents, and imitation. *RAND Journal of Economics*, **40**, 611–635.

Bessen, J. and M. J. Meurer (2008) Do patents perform like property? *Academy of Management Perspectives*, **22**, 8–20.

Boldrin, M. and D. K. Levine (2008) *Against Intellectual Monopoly*. New York: Cambridge University Press.

Braudel, F. (1985) *La dynamique du capitalisme*. Paris: Flammarion.

Guellec, D., T. Madiès and J.-C. Prager (2010) *Les marchés de brevets dans l'économie de la connaissance*. Paris: Conseil d'analyse économique, La Documentation Française.

Hall, B. H. (2007) Patents and patent policy. *Oxford Review of Economic Policy*, **23**, 1–20.

Harhoff, D., G. von Graevenitz and S. Wagner (2008) *Incidence and Growth of Patent Thickets: The Impact of Technological Opportunities and Complexity*, CEPR Discussion Paper 6900. London: Centre for Economic Policy Research.

Hunt, R. M. (2006) When do more patents reduce R&D? *American Economic Review*, **96**, 87–91.

Kyle, M. and A. McGahan (2009) *Investments in Pharmaceuticals Before and After TRIPS*, NBER Working Paper 15468. Cambridge, MA: National Bureau of Economic Research.

Lester, R. K. (2005) *Universities, Innovation and the Competitiveness of Local Economies: A Summary Report from the Local Innovation Systems Project*, Massachusetts Institute of Technology Industrial Performance Center Working Paper 05–010. Cambridge, MA: MIT.

Leydesdorff, L. and M. Meyer (2009) The decline of university patenting and the end of the Bayh–Dole effect. *Scientometrics*, **83**, 355–362.

Machlup, F. (1958) *An Economic Review of the Patent System*. Washington, DC: US Government Printing Office.

Penrose, E. (1951) *The Economics of the International Patent System*. Baltimore, MD: John Hopkins University Press.

Scotchmer, S. (2004) *Innovation and Incentives*. Cambridge, MA: MIT Press.

Index